AF181577

# Leptospirosis - Symptoms, Causes and Treatment

*Edited by Sara Savić and Marina Žekić*

Published in London, United Kingdom

Leptospirosis - Symptoms, Causes and Treatment
http://dx.doi.org/10.5772/intechopen.111184
Edited by Sara Savić and Marina Žekić

Contributors
Abdolreza Babamahmoodi, Aregitu Mekuriaw, Arfan Ali, Beatriz Rivas Sánchez, Chandrani Goswami, Claudia Ferreira, Durlav Prasad Bora, Farhang Babamahmoodi, Joselito S. Balingit, Kalimuthusamy Natarajaseenivasan, Marie-Françoise Doursout, Marina Žekić, Nagendra Nath Barman, Nur Abdul Kader, Patrick Banza, Pooja Shivshankar, Sara Savić, Sophia Makdoh Gogoi, Yann A. Meunier

© The Editor(s) and the Author(s) 2025
The rights of the editor(s) and the author(s) have been asserted in accordance with the Copyright, Designs and Patents Act 1988. All rights to the book as a whole are reserved by INTECHOPEN LIMITED. The book as a whole (compilation) cannot be reproduced, distributed or used for commercial or non-commercial purposes without INTECHOPEN LIMITED's written permission. Enquiries concerning the use of the book should be directed to INTECHOPEN LIMITED rights and permissions department (permissions@intechopen.com).
Violations are liable to prosecution under the governing Copyright Law.

Individual chapters of this publication are distributed under the terms of the Creative Commons Attribution 3.0 Unported License which permits commercial use, distribution and reproduction of the individual chapters, provided the original author(s) and source publication are appropriately acknowledged. If so indicated, certain images may not be included under the Creative Commons license. In such cases users will need to obtain permission from the license holder to reproduce the material. More details and guidelines concerning content reuse and adaptation can be found at http://www.intechopen.com/copyright-policy.html.

Notice
Statements and opinions expressed in the chapters are these of the individual contributors and not necessarily those of the editors or publisher. No responsibility is accepted for the accuracy of information contained in the published chapters. The publisher assumes no responsibility for any damage or injury to persons or property arising out of the use of any materials, instructions, methods or ideas contained in the book.

First published in London, United Kingdom, 2025 by IntechOpen
IntechOpen is the global imprint of INTECHOPEN LIMITED, registered in England and Wales, registration number: 11086078, 167-169 Great Portland Street, London, W1W 5PF, United Kingdom

For EU product safety concerns: IN TECH d.o.o., Prolaz Marije Krucifikse Kozulić 3, 51000 Rijeka, Croatia, info@intechopen.com or visit our website at intechopen.com.

British Library Cataloguing-in-Publication Data
A catalogue record for this book is available from the British Library

Leptospirosis - Symptoms, Causes and Treatment
Edited by Sara Savić and Marina Žekić
p. cm.

This title is part of the Infectious Diseases Book Series, Volume 37
Topic: Bacterial Infectious Diseases
Series Editor: Alfonso J. Rodriguez-Morales
Topic Editors: Tomas Jarzembowski and Katarzyna Garbacz

Print ISBN 978-0-85466-824-3
Online ISBN 978-0-85466-823-6
eBook (PDF) ISBN 978-0-85466-825-0
ISSN 2631-6188

If disposing of this product, please recycle the paper responsibly.

IntechOpen
intechopen.com

Built by scientists, for scientists

Explore all IntechOpen books

IntechOpen Book Series

# Infectious Diseases

Volume 37

## Aims and Scope of the Series

This series will provide a comprehensive overview of recent research trends in various Infectious Diseases (as per the most recent Baltimore classification). Topics will include general overviews of infections, immunopathology, diagnosis, treatment, epidemiology, etiology, and current clinical recommendations for managing infectious diseases. Ongoing issues, recent advances, and future diagnostic approaches and therapeutic strategies will also be discussed. This book series will focus on various aspects and properties of infectious diseases whose deep understanding is essential for safeguarding the human race from losing resources and economies due to pathogens.

# Meet the Series Editor

Dr. Rodriguez-Morales is a Senior Researcher and Faculty at the Faculty of Medicine of Fundación Universitaria Autónoma de las Américas-Institución Universitaria Visión de las Américas, Pereira, Risaralda, Colombia. He is also a Researcher and Faculty at the Faculty of Health Sciences of the Universidad Científica del Sur, Lima, Peru. He was Non-Resident Faculty Researcher of the Gilbert and Rose-Marie Chagoury School of Medicine, Lebanese American University, Beirut, Lebanon (2023-2024). Dr. Rodriguez-Morales is an expert in tropical and emerging diseases, particularly in zoonotic and vector-borne diseases, including COVID-19 and mpox, and their occurrence among travelers and migrants, as well as its approach under the One Health lens. He is currently the President of the Latin American Society for Travel Medicine (SLAMVI) (2023-2025). Past-President, Colombian Association of Infectious Diseases (ACIN) (2021-2023). Member, Committee on Tropical Medicine, Zoonoses, and Travel Medicine, ACIN. Member of the Council, International Society for Infectious Diseases (ISID) (2020-2026). Editor-in-Chief, Travel Medicine and Infectious Diseases. Full Member of the European Society for Clinical Microbiology and Infectious Diseases (ESCMID). Member of the American Society for Tropical Medicine & Hygiene (ASTM&H). Member of the International Society for Travel Medicine (ISTM). He was a member of the Technical Working Group of the World Health Organization (WHO) for the Global Research Agenda on Health and Migration and the 3rd Global Consultation on the Health of Migrants and Refugees (co-organized by WHO, UNHCR and IOM) (2023). Currently, Member of the World Health Organization (WHO) Guideline Development Group for Clinical Management of post-COVID-19 condition (2024-2025). H index 66 (Scopus) with 867 documents indexed in Scopus.

# Meet the Volume Editors

Sara Savić, Ph.D., DVM, is a senior researcher fellow and the Head of the Department for Serology, Immunology, and Biochemistry at the Scientific Veterinary Institute "Novi Sad" in Novi Sad, Serbia. She is also head of the National Reference Laboratory for leptospirosis, Q fever and Maedi-Visna disease in veterinary medicine. Her primary work focuses on diagnostic procedures for zoonotic diseases, primarily in animals. Dr. Savić completed her Ph.D. degree in diagnostics of Lyme disease in ticks and dogs, and then turned her interests towards other zoonotic diseases important for public health (such as brucellosis, leptospirosis, Q fever, etc.), including vector-borne diseases of bacterial and parasitic origin (such as dirofilariasis, leishmaniasis, anaplasmosis, ehrlichiosis, etc.). Today, she examines these diseases from a One Health perspective and is committed to addressing zoonotic diseases from a One Health approach. Dr. Savić is an active member of the Network for Ecohealth and One Health, as well as the Serbian Association for One Health.

Marina Žekić, Ph.D., DVM, is a senior research associate at the Department for Serology, Immunology and Biochemistry in the Scientific Veterinary Institute "Novi Sad" in Novi Sad, Serbia. Her primary work focuses on molecular diagnostic procedures for detecting zoonotic pathogens, primarily in animal samples. After completing her Ph.D., she started working on molecular techniques for diagnostic use in veterinary medicine, which is important for public health. She is a researcher looking forward to new and advanced technologies and implementing them in laboratory work. She is also involved in the National Reference Laboratory in Veterinary Medicine's work to diagnose leptospirosis.

# Contents

# Preface

Leptospirosis is an infectious disease that has been globally present for centuries, but it is still one of the overlooked threats to global health. It is a zoonotic disease affecting humans and animals of many species, representing a great burden in some regions. Leptospirosis has a broad geographic distribution and considerable morbidity and mortality, but still, this disease is sometimes underdiagnosed and therefore also underreported. The disease has a nonspecific clinical presentation and some diagnostic challenges, but lack of general awareness is probably the most important factor.

The book represents an effort to analyze current knowledge on leptospirosis, bringing together perspectives of human and veterinary medicine, environmental health, microbiology and public policy. The One Health approach is essential for understanding the epidemiology and risk factors related to leptospirosis.

The introductory chapter slowly takes the readers into the topic of leptospirosis, with a few overall information about the disease.

The first chapter is about the epidemiology of leptospirosis in the broadest possible way. It covers the infection cycle and transmission of the pathogen, pathogenesis, diagnostic, pathogenic and biological characteristics of *Leptospira* spp. It also covers diagnostic challenges, reservoirs of the pathogen and different factors that can impact the outbreaks.

Chapter two presents the clinical landscape of leptospirosis, highlighting the variability in clinical symptoms and difficulties of early diagnosis. The concept of chronic leptospirosis is also presented and needs future research in diagnosis and long-term management.

Chapter three focuses on the epidemiology and public health dimension of leptospirosis. Outbreaks of leptospirosis often follow seasonal patterns and environmental situations such as floods or heavy rainfalls. Rodents are usually identified as pathogen reservoirs, and the risk for infection rises with increased exposure to contaminated water or soil. The chapter also highlights the importance of improved diagnostic tools, particularly in regions where resources are limited and where the disease is most prevalent.

The fourth chapter concerns animal leptospirosis, underlining the economic and ecological impact. In veterinary medicine, leptospirosis can cause reproductive failure, a decrease in milk production, and livestock mortality, all leading to significant economic losses. Small mammals and rodents are also mentioned as reservoirs of the disease, with the possibility of excreting *Leptospira* in the urine without showing any symptoms of the disease, continuously contaminating the environment. Therefore, surveillance of the disease goes much beyond only the human population and must also include animal health monitoring and biosecurity measures.

Chapter five explains the relationship between leptospirosis and environmental conditions, particularly in tropical and subtropical climates that are favourable for the survival of *Leptospira*. Warm temperatures, high humidity, and still or slow-moving water contribute to the maintenance of the pathogen. Urbanization and land use changes also interfere with ecological balance, increasing human/animal contact with the possibility of pathogen transmission. The chapter calls for environmental stewardship and cross-sector collaboration.

The sixth chapter looks into the global burden of leptospirosis and policy level responses to the outbreaks. Leptospirosis has been acknowledged by international organizations as a neglected tropical disease with epidemic potential. There is a need to enhance the surveillance system and educational campaigns for populations at risk.

Besides the biological and clinical complexity of leptospirosis, there is also the implication of global health equity. The disease affects vulnerable populations living in poor hygienic conditions, poverty, flood areas, and often in close contact with rodents. All of this highlights the intersection between health, environment and social determinants. Addressing leptospirosis is not only a medical question, but a holistic approach is also required, considering environmental factors.

This book is intended for a broad and diverse audience who wants to know about leptospirosis. Clinical insights relevant to diagnosis and management are shown. There is also some epidemiological and policy information relevant to public health. Some animal health implications are also mentioned from the veterinary medicine point of view. There is also a section with ecological dynamics and an environmental approach. On the other hand, there is much more information about leptospirosis than covered in this book, because it is an excellent research field with many approaches.

The main purpose of this book is to create awareness, curiosity, and debate about leptospirosis, as it is a complex disease that has been present for a long time and is still causing outbreaks worldwide. The awareness of leptospirosis is meant to be raised as an often-neglected component of global health. Curiosity and debate are meant to be stimulated between different disciplines, resulting in stimulating interdisciplinary dialogue and activities, leading further to a multisectoral approach combining the strengths of human and veterinary medicine, environmental research and public policy.

Our current world is shaped by ecological disruption, rapid urbanization, global interconnectivity, and threats of zoonotic diseases. Therefore, preparedness for the future is needed, and it lies in knowledge. This book is meant to contribute to that knowledge.

**Sara Savić and Marina Žekić**
Scientific Veterinary Institute “Novi Sad”,
Novi Sad, Serbia

Chapter 1

# Introductory Chapter: Leptospirosis as a Global Concern at a Glance

*Sara Savić and Marina Žekić*

## 1. Introduction

Leptospirosis is a bacterial zoonotic disease and is one of the most prevalent ones in the world. This disease is a significant public burden not only in tropical and subtropical regions but also in some other parts of the world. Despite the worldwide presence, many people are unaware of the risks, symptoms, and prevention activities. Understanding of leptospirosis has several aspects, such as biology, transmission, clinical manifestation, diagnostics, treatment and prevention.

Leptospirosis is an infectious disease caused by the bacterium Leptospira. This bacterium mostly infects animals, but it can also affect humans. The infectious agent can be found in domestic and wild animals, such as rodents, dogs, and humans. Bacteria Leptospira are mostly present in and shed by the urine of infected animals, and that way they can contaminate water, soil or food. Humans are mostly infected by contact with the environment contaminated with the bacteria. For example, contact with contaminated water through broken skin, such as cuts or scratches on humans or animals, can provide a pathway for bacteria to enter the body.

Bacteria can survive in water or soil for some time, especially in warm and humid conditions, which makes Leptospira a resilient microorganism and therefore represents a global concern.

The first known record of leptospirosis was in China, identified in rice harvesters, and in Japan, it was named "autumn fever." A disease characterized by jaundice and renal failure in a severe form was described by Adolf Weil in 1886, so the disease was named at the time Weil's disease. In 1907, Arthur Stimson identified the causative agent of leptospirosis by finding the spirochetes in the kidneys of the patient and named it Spirochaeta interrogans because of the question mark shape [1]. The rats as carriers of the disease were identified in 1917 and then also as reservoirs of leptospirosis [2]. Different serovars of Leptospira were identified much later, and by the 1950s, more than 200 serovars were found, and some were already associated with a specific animal as a host [1]. During the past few decades, leptospirosis has been found to be a re-emerging disease related to environmental and climate changes. When first discovered, leptospirosis was declared as an occupational disease. Still, today, it is a global health concern that affects animals and humans.

Leptospirosis can be found in any part of the world, but its highest prevalence is still in tropical and subtropical regions, such as Southeast Asia, Central and South America, and some parts of Africa. A warm and humid climate and frequent flooding are all good preconditions for both the spread and survival of Leptospira pathogens. The factors that can contribute to spreading the disease in urban areas are poor hygienic conditions, which favor rodents' reproduction and close contact with humans. Rodents are primary carriers of Leptospira pathogens, and they excrete them with their urine into water, soil or other surfaces, contaminating them. Outbreaks of leptospirosis are often related to natural disasters, most often floods [3]. Floods represent a risk factor for the spread of leptospirosis due to the ability of Leptospira to survive in water or humid places for a longer time than in dry places [4, 5].

Leptospirosis is an endemic disease in many parts of the world, and also obligative for reporting, but even so, it is often underreported. That might be because of the nonspecific symptoms it causes in clinical cases, or in some countries, there might not be the possibility of adequate diagnostic testing [4]. On the other hand, in developed countries such as the US, leptospirosis is less spread and less of a concern, but it is still recognized and associated with some risk factors, such as outdoor activities or occupational factors related to animals.

Animals of different species can be affected by leptospirosis, including pets and domestic and wild animals. Canine leptospirosis can become a major concern because of its close connection to humans, but there is also a vaccine for dogs that is applicable in many countries. In livestock and farm animals, leptospira outbreaks can happen most frequently in cattle, pigs, and sheep. The disease causes reproductive problems such as abortion, kidney damage and, after all, decreased productivity. Wildlife animals, including rodents, can also represent a reservoir of Leptospira pathogens which can be spilled into the environment from which humans and other animals are affected.

One Health approach recognizes the interconnections between humans, animals, and the environment, and leptospirosis is an ideal case for implementing this approach. Leptospira pathogens can be transmitted from animals, such as rodents, livestock, or even pets, to humans through contaminated water, soil, or animal excrement. This transmission pathway highlights all three components of the One Health approach. So, addressing leptospirosis requires an interdisciplinary engagement in order to monitor the animals as reservoirs, control the rodent population, raise awareness in humans, etc. Intraspecies and interspecies transmission of Leptospira pathogens depend on the reservoir host animal where bacteria replicate and shed in urine, the persistence of bacteria in the environment and the interaction between humans, animals and the environment [6]. One Health approach in the case of leptospirosis outbreaks, therefore, involves veterinarians, medical doctors, public health officials, environmental scientists, biologists, etc. The One Health approach aims not only to control outbreaks but also to prevent future occurrences by addressing all aspects of the disease. A multidisciplinary team should focus on building preparedness and resilience, particularly in endemic regions where the disease remains a constant threat.

## 2. Conclusion

Leptospirosis continues to be a significant global health issue and concern. Its complex nature involves environmental, animal, and human factors that contribute to the transmission of the pathogen and the spreading of the disease. The main transmission of the pathogen goes through contaminated water, soil, or animals such as rodents, dogs, livestock, and potentially wildlife. The impact of the disease is the most severe in tropical and subtropical regions, especially in places with a high population density, poor sanitation, and frequent flooding, which create perfect conditions for the causative pathogen to survive and spread.

Misdiagnoses and underreporting need to be improved globally. Advanced diagnostic methods should improve the ability to perform quick and precise detection of the Leptospira pathogen. The efforts should be made to invest people's resources and finances into surveillance and monitoring activities, especially in regions with high prevalence. This way, the places with high exposure can be identified and marked. A One Health approach with mutual efforts between human, animal and environmental sectors is essential for leptospirosis control. The strategy of One Health can lead to more effective prevention and control measures in regions where an interdisciplinary approach is implemented, meaning rodent control, vaccination, surveillance and education.

## Author details

Sara Savić* and Marina Žekić
Scientific Veterinary Institute "Novi Sad", Novi Sad, Serbia

*Address all correspondence to: sara@niv.ns.ac.rs

IntechOpen

© 2025 The Author(s). Licensee IntechOpen. This chapter is distributed under the terms of the Creative Commons Attribution License (http://creativecommons.org/licenses/by/3.0), which permits unrestricted use, distribution, and reproduction in any medium, provided the original work is properly cited.

## References

[1] Adler B. History of leptospirosis and leptospira. Current Topics in Microbiology and Immunology. 2015;**387**:1-9. DOI: 10.1007/978-3-662-45059-8_1

[2] Levett PN. Leptospirosis. Clinical Microbiology Reviews. 2001;**14**(2):296-326. DOI: 10.1128/CMR.14.2.296-326.2001

[3] Pappas G, Papadimitriou P, Siozopoulou V, Christou L, Akritidis N. The globalization of leptospirosis: Worldwide incidence trends. International Journal of Infectious Diseases. 2008;**12**:351-357. DOI: 10.1016/j.ijid.2007.09.011

[4] Costa F, Hagan JE, Calcagno J, Kane M, Torgerson P, Martinez-Silveira MS, et al. Global morbidity and mortality of leptospirosis: A systematic review. PLoS Neglected Tropical Diseases. 2015;**9**(9):e0003898. DOI: 10.1371/journal.pntd.0003898

[5] Naing C, Reid SA, Aye SN, Htet NH, Ambu S. Risk factors for human leptospirosis following flooding: A meta-analysis of observational studies. PLoS One. 29 May 2019;**14**(5):e0217643. DOI: 10.1371/journal.pone.0217643

[6] Sykes JE, Haake DA, Gamage CD, Mills WZ, Nally JE. A global one health perspective on leptospirosis in humans and animals. Journal of the American Veterinary Medical Association. 2022;**260**(13):1589-1596. DOI: 10.2460/javma.22.06.0258

## Chapter 2

# Comprehensive Epidemiological Insights on Leptospirosis Global Health Effects and the Risk Factors for Prevention and Control

*Aregitu Mekuriaw*


## Abstract

Leptospirosis is a potentially fatal and often ignored zoonotic illness caused by the genus Leptospira. The pathogen infects humans, animals, and ecosystems with diverse clinical signs and is estimated to be responsible for 60,000 mortalities, with over a million cases annually. It is common and poses a significant diagnostic problem in low-income tropical and subtropical countries. The incidence is seasonal; climate change, animal exposure, physical environment, and globalization are risk factors of leptospirosis. Rats are the primary reservoir species, while other mammals may potentially play a major role in the transmission of human illnesses. The high burden of pathogens on animals affects the livestock reproduction sector and impacts the socioeconomic burden. Human leptospirosis commercial vaccines are available in Japan, China, Cuba, and France. The pathogen's ability to exist in the environment contributes to its growth in water and enables the maintenance and distribution of the bacteria globally. This situation reflects a higher impact on agriculture, human and animal health, and ecosystems, which need comprehensive management and a holistic approach, promoting different disciplines and joint-work initiatives toward controlling and treating leptospirosis. Therefore, this book chapter is aimed at highlighting the leptospirosis disease epidemiology, its global health effects, the risk factors, and the challenges in leptospirosis disease control and prevention.


**Keywords:** epidemiology, global, leptospirosis, prevention, zoonosis

## 1. Introduction

Leptospirosis is the most globally prevalent neglected zoonosis disease that affects both domestic and wild animals, as well as people mainly in low-income populations living in both urban and rural areas of tropical and subtropical regions

[1, 2]. Historically, leptospirosis has been linked to occupational threats, upsetting soldiers [3], farmers [4, 5], miners [6], and abattoir workers [7], but individuals in non-specific occupational groups may also get leptospirosis. Thus, with a global distribution, leptospirosis concerns human health and safety at work due to direct or indirect contact with the urine of infected animals and through contaminated water. It can result in life-threatening clinical issues like pulmonary hemorrhage syndrome, which has a high morbidity and mortality rate [8]. Despite having a significant negative impact on human health and livestock production, leptospirosis is a tropical disease that is frequently neglected [9]. Leptospirosis is a zoonosis that has a significant global mortality and morbidity rate and impacts the interface between humans, animals, and ecosystems with several clinical symptoms. It is instigated by the pathogenic spirochete of the genus *Leptospira*. There is evidence that *Leptospira* can be found in the tissues of almost all mammal species, and even reptiles and birds may be susceptible to them [10, 11]. Most of the time, an infection has no symptoms at all and is mild, generic, or indiscriminate. But rarely, leptospirosis can cause severe symptoms in both humans and animals that can be life-threatening clinical issues, such as pulmonary hemorrhage syndrome, which has a high morbidity and mortality rate [8, 12].

It is estimated that leptospirosis is responsible for 60,000 deaths annually, with over a million cases reported [13, 14]. However, due to low clinical presentation specificity, lack of diagnostic facility availability, low reliability of surveillance systems, and little knowledge of the disease, these numbers are thought to be significantly underestimated [15]. Effective vaccination is the sole way to avoid it, but since the discovery of bacteria more than a century ago, the leptospirosis vaccine has not advanced significantly [16, 17]. Numerous investigations have revealed that wild animals such as rats, bats, foxes, sea lions, and capybaras are leptospirosis reservoirs [18–20] and may be spread to people by wild boars, monkeys, and tenrecs [21–23]. Thus, the identification of animal hosts, which are both direct and indirect infection spreaders, is an essential first step in managing and controlling leptospirosis, regardless of the approach employed, as these animals are the ones that discharge the bacteria into the environment and cause environmental contamination [24]. Among many other considerations, the animal population, environmental circumstances, infectious serovars, accessibility to animals, and species involved can all influence the choice of treatments [25]. Furthermore, leptospirosis still presents several obstacles in underdeveloped nations, encompassing not only public health domains but also medical and biological diagnostics, as well as case management [26]. With varying clinical symptoms, this pathogen affects people, domestic animals, wildlife, and ecosystems. Although, the interplays between domestic animals, wildlife, and humans can occur in a range of locations and are associated with significant epidemics around the globe [27, 28].

Urine from carrier animals can infect humans through direct contact with skin lesions or mucous membranes, while contact with fresh water bodies or irrigated soil polluted by animal urine is more common [14, 17]. Leptospirosis is more frequently associated with leisure and occupational activities in temperate areas; on the other hand, as a result of increased exposure to polluted soil or water because of high rodent populations, the majority of outbreaks in tropical countries happen during the rainy season. More than 35 pathogenic and intermediate *Leptospira* species, which are mostly spread by environmental factors, are the cause of human leptospirosis [29]. In poor nations from tropical and subtropical locations across the world, leptospirosis has a severe negative impact on vulnerable farming communities. Most animal infections occur through contact with the environment, while some mammal species can frequently contract infections through intercourse. However, the prevalence of infections in animals caused by various *Leptospira* serovars on livestock farms in tropical and subtropical countries is unknown, which accounts for the lack of standardization and accessibility to diagnostic testing [30]. The high burden of the pathogen on animal's health affects livestock, the high death rate, the reproduction sector, and the socioeconomic burden [9]. Additionally, they have an impact on the environment because of their widespread distribution and the fact that water promotes the growth, maintenance, transmission, and dispersal of bacteria [31]. Government agencies in charge of leptospirosis treatment and control are promoting a holistic approach, which emphasizes the necessity for inclusive management given the influence on agriculture, human, and animal health [32].

Transmission in vulnerable species is accelerated by epidemiological risk factors, primarily in endemic locations, and includes leisure activities, travel abroad, and occupational exposure. Currently, the way in which *Leptospira* exposure occurs has changed due to tourism and globalization. As per recent studies [33–36], leptospirosis is currently seen as a significant danger for travellers. Outdoor recreation and sports have been connected to leptospirosis [37, 38]. Additionally, from the standpoint of individual health, the commercialization of domestic and wild animals in non-endemic regions is a significant factor in the transmission of leptospirosis [9]. Therefore, the discovery of Leptospira reservoirs in wild species provides new information about the epidemiology of the pathogen. Numerous studies have reported that a wide range of serious infectious diseases, such as hanta virus pulmonary syndrome, dengue fever, yellow fever, malaria, trypanosomiasis, leishmaniasis, and leptospirosis, are caused by local climate change, large-scale deforestation, disruption of significant ecosystems and ecosystem services, and urbanization [28]. Furthermore, leptospirosis is an issue for animal health and a source of economic loss in the same underdeveloped areas where the burden of human disease is high; it is acquiescent to One Health approach to intervention [39]. In general, the impact of pathogen backgrounds on agriculture, human health, and animal health discloses the need for inclusive management, where a holistic approach is promoted for controlling and treating leptospirosis. The inclusive management of leptospirosis necessitates a deeper comprehension of the illness and the biological, socioeconomic, and cultural risk factors present in the area. Additionally, as zoonoses are now a growing public health concern, it is critical to support collaborative work projects and provide proof of the necessity for work from a One Health perspective. In this chapter presents an updated, comprehensive epidemiological insight on leptospirosis global health effects and the risk factors for prevention and control.

### 1.1 Leptospirosis global zoonotic health effects

A zoonotic bacterial disease, leptospirosis, affects populations with few resources the most, while it can arise in a variety of epidemiological contexts [40, 41]. Because a wide variety of mammals can retain and excrete the spirochete agent from their renal tubules, the disease has a wide geographic spread [41, 42]. Leptospirosis is predicted to be responsible for more than half a million deaths, with over a million cases reported annually [13, 14]. Costa et al.'s [41] more recent estimates indicate that leptospirosis is "among the leading zoonotic causes of morbidity and mortality". However, according to estimates from the World Health Organization (WHO), there are five endemic cases of human leptospirosis for every 100,000 persons and 14 epidemic cases each year worldwide [43]. These disparate figures demonstrate how challenging it is to quantify the current leptospirosis burden worldwide. This is due to the lack of diagnostic facility availability, low reliability of surveillance systems and little knowledge of the disease, and these numbers are thought to be significantly underestimated [15]. Furthermore, the diagnostic tests that are currently accessible are not ideal; the best tests miss nearly 40% of patients [15]. However, due to low clinical presentation specificity and close similarities to other diseases, only one in ten infections are believed to be precisely identified [44]. Case fatality rates for leptospirosis can vary from less than 5% to more than 30%, depending on the clinical presentation and course of treatment. Young adults between the ages of 20 and 49 are more likely to contract an infection with *Leptospira* [41]. Males are estimated to bear 80% of the global leptospirosis burden [45]. Leptospirosis, which causes high death rates in both people and sensitive animal species, has become a major infectious zoonotic disease in the last 10 years, especially in endemic tropical locations with significant rainfall [14, 46]. The high death rate in leptospirosis is still high, attributed to a variety of poorly understood factors that may include the innate pathogenicity of some *Leptospira* strains or host immune-pathological responses that are genetically determined [47]. These factors also contribute to delays in diagnosis brought on by a lack of infrastructure and sufficient clinical suspicion.

In many areas where transmission is endemic, the primary burden attributed to leptospirosis life-threatening manifestations has occurred as a significant cause of acute renal injury related to Weil's disease [41] and pulmonary hemorrhage syndrome [48]. Moreover, leptospirosis is becoming more widely acknowledged as a significant contributor to undifferentiated fever [49, 50]. Most cases of leptospirosis are misinterpreted as dengue [50, 51], or other acute febrile illness-related conditions. This has been linked to the lack of an appropriate diagnostic test [52], which leads to underreporting of cases [53, 54] and deaths [51]. The absence of trustworthy leptospirosis burden approximations has hindered attempts to build the case for investment in order to remove important obstacles, such as better diagnostics, and find efficient preventive and control strategies. Risk groups for leptospirosis include those working in sewage and abattoirs, military personnel, and people who participate in water sports and recreation, as well as those exposed to animal reservoirs or contaminated surroundings [41]. Furthermore, because of the effects of globalization and climate change, leptospirosis has become a health concern in previously unexplored areas [55, 56]. Epidemics are now known to be glowed by disasters and harsh weather [40, 41].

### 1.2 Socioeconomic impact of leptospirosis

In addition to the health consequences that leptospirosis might have, the relatives of the victims and society at large may also be negatively impacted financially and socially. The condition has both direct and indirect financial consequences, such as the expense of treating the acute sickness, managing long-term medical issues, causing income loss due to the illness, and maybe having an impact on one's eventual ability to earn a living [57]. During epidemics, thousands of individuals may possibly become infected in a short period of time, placing a great deal of strain on healthcare services. The health system will be further taxed by the implementation of public health initiatives for surveillance, control, and prevention. Moreover, leptospirosis poses a risk to cattle, exacerbating financial losses. A lot of leptospirosis epidemics happen in urban slums, which are home to some of the most vulnerable and impoverished individuals on the planet. These areas also tend to have low socioeconomic resilience, insufficient access to healthcare, and few public health resources. The combination of these factors will make leptospirosis the most harmful disease to communities in terms of both health and socioeconomics, and as emerging nations become more urbanized, this situation is expected to get worse.

### 1.3 Leptospirosis epidemiological insights

Leptospirosis is a zoonotic, important bacterial disease that is associated with low levels of sanitation and poverty in both rural and urban regions. Although transmission happens in both developed and developing nations, the occurrence of human infection is greater in tropical regions than in temperate ones. Since the disease is not well-known and tests are not readily available or performed quickly enough, incidence rates are underestimated. In endemic areas, infection that is asymptomatic or subclinical is frequent [58]. It is also interconnected with the locations where livestock is raised, as close contact with animals may facilitate the transmission of leptospirosis. Veterinary professionals, hunters, caregivers of wild animals, farmers, and agricultural workers who come into contact with infected animals are among the people who are affected by this occupational risk disease [59]. *Leptospira* strains associated with human cases are influenced by the animal reservoir in each particular area. Numerous mammals, such as domestic animals, can preserve and disperse a variety of *Leptospira* serovars in rural areas [60, 61].

Due to the fact that numerous species can act as reservoirs for illness and are found in tropical, subtropical, and temperate regions, leptospirosis has a complicated epidemiology [62]. Both genders and urban and rural environments report cases of the disease, with young adult males being the most affected [44, 63, 64]. Additionally, it is a significant occupational zoonosis that affects workers in the agricultural, sewage, mining, slaughterhouse, butcher, dairy, veterinary, and animal handling industries, as well as construction, kennel, military, and fishing industries [63–65]. Teams that work in occupationally exposed environments consider communities living in urban slums with inadequate sanitation to be extremely vulnerable and represent a substantial risk to occupationally exposed personnel [66–69].

Leptospirosis was mostly a work-related illness in temperate, wealthy nations, linked to exposure to animals or freshwater. It is becoming more widely acknowledged that this environmental illness is linked to leisure pursuits including tramping, rafting, and other outdoor sports [60, 70]. There is evidence that the countries visited pose harm to the environment, and the number of individuals with leptospirosis cases returning from tropical vacations is rising [34, 71]. Leptospirosis is known to occur in both temperate and tropical climates. Research has revealed that the highest frequency of the illness transpired during the rainy seasons in tropical areas and late summer to early fall in temperate countries; nevertheless, the majority of the outbreaks were preceded by times of excessive rainfall where favorable conditions allow for its spread [44, 72, 73]. This is due to warm and moist weather which favors *Leptospira* existence, and incidences in tropical areas are almost ten times greater than those in temperate areas.

Both global warming and changes in rainfall patterns have the potential to raise the disease burden. For instance, it is a seasonal sickness that is most likely caused by climate change environment (soil or water) contamination in particular as a result of prolonged exposure to an abundance of rodent populations [74–77]. In areas of moderate climate, work-related and entertaining activities (e.g. contact with polluted water or soil, contact with cattle infected with leptospirosis) are more probable to be linked with leptospirosis [77, 78]. Thus, the population biology, behaviour, or community ecology of spirochetes and their hosts are altered by environmental factors, which have a significant impact on leptospirosis transmission.

## 1.4 Aetiology of leptospirosis

*Leptospira* are tiny, helicoidal bacteria with a diameter of 0.1–0.2 μm and a length of 6–20 μm. Two periplasmic flagella are located at each pole of these hooked-ended, gram-negative spirochete bacteria. In addition to aiding in the bacterium's survival within the host and during infection, flagella allow the bacteria to travel swiftly in viscous environments such as blood, interstitial fluid, and connective tissues [17]. Lipo-polysaccharides, proteins, lipoproteins, and lipids make up the outer membrane, while lipoproteins and transport proteins are in the double-layer inner membrane, which is tightly linked to the peptidoglycan layer [79]. It is believed that having periplasmic flagella and a corkscrew-shaped structure helps *Leptospira* survive over the long term in both their host and the environment [80]. *Leptospira* were categorized formerly based on their antigenic patterns, which were grouped into over 25 serogroups that combined serovars that were antigenically linked [58, 81], but, in more recent times, a molecular taxonomy has been reported that uses DNA relatedness to split the *Leptospira* genus into many species [82]. *Leptospira* is classified as pathogen, intermediate, or saprophyte strain. Saprophytes are normal inhabitants of many types of wet, humid, and freshwater environments. Pathogen strains can be found in nature in the renal tubules of animals and are notable for their extreme sensitivity to UV radiation, chlorine, and detergents [60, 83]. It is also believed that they are vulnerable to low temperatures and acidic environments.

Almost all mammal species are susceptible to *Leptospira* infection, which can result in leptospirosis. Intermediates strains can infect mammals, but they often have a lesser virulence and only cause minor illnesses. Pathogenic strains can endure extended periods of time in the environment, but their capacity to multiply outside of their live host is limited. As a response to the prolonged lack of a suitable host species, they instead occasionally display binary fission or, in other situations, group together to form a biofilm [84]. For survival, both saprophyt and pathogen *Leptospira* need the same kinds of nutrients and temperatures. Saprophyte strains are sometimes discovered in humans and other animals, but they usually live their whole lives in soil and water. In order to finish their life cycles, pathogenic organisms need a live object. However, *leptospires'* ability to survive and thrive is mostly dependent on the temperature, pH, and moisture content of their surroundings, in addition to the vital nutrients they need to grow [85, 86]. *Leptospira* are assumed to be persistent and have an extended lifespan in the environment because of their ability to build biofilms with other bacteria [86].

The majority of settlements use natural water sources like rivers, streams, or subterranean aquifers to gather water, which is then stored in reservoirs for extended periods of time. It is reported worldwide as a threat to epidemics in developing nations during and after the rainy season. People in metropolitan areas are susceptible to leptospirosis when roadways flood after heavy rainfall [87, 88]. There is a significant risk of infection for farmers in rural areas who labour in rice fields [89].

### 1.5 Leptospirosis infection cycle and transmission

The possibility of contracting leptospirosis is *via* contact with contaminated environment, infected wild animals, and rodents [60, 61, 68]. Natural reservoirs for *Leptospira*, such as herbivores and other animals, may develop into symptomless carriers following an infection. As a result, they serve as a significant foundation for *Leptospira* in the environment through bacteria emissions in urine. Renal carrier state is thought to be the primary mechanism that allows leptospirosis to sustain itself in the environment [90]. The primary carriers of *Leptospira* are rodents, who can contaminate the environment by releasing the germs through their urine [17]. Many wild and domestic animals, such as bovine, swine, canines, and horses, can serve as reservoirs and contribute to the spread of leptospirosis, but the brown rat has been shown to be the most vital source of human infection [24]. Humans can get sick from direct contact with infected animals and *Leptospira* contaminated soil and water. Once ingested, the germs penetrate and enter the bloodstream and go to the organs, where they cause pathological harm.

In the past, leptospirosis was mostly thought to be an occupational illness linked to occupations including mining, sewage cleaning, raising and killing livestock, veterinary care, and military operations. With the introduction of preventive measures, the relative significance of certain occupational risks has dropped. Human infection occurs when a person comes into contact with a carrier mammal's contaminated urine, either directly or through polluted surfaces, water, or soil [91]. Leptospirosis is primarily spread by damaged skin or mucosa when contaminated water, soil, or urine comes into contact with it [92]. *Leptospira* has never been shown to represent significant epidemiological sources of transmission, despite the fact that humans can excrete them into their urine for weeks, or, in rare cases, months, or longer than a year [24, 60]. In addition, the occurrence of different *Leptospira* serovars within a human population is influenced by the existence of reservoir animals and the serovars they carry, as well as by local environmental conditions, occupation, agronomical practices, and agricultural methods. People who live in urban areas that are characterized by high population density, poor housing quality, and inadequate cleanliness are more likely to be exposed to rats and consequently contract leptospirosis. In industrialized countries, occupational exposure, visits to endemic regions, leisure activities, and the introduction of both domestic and wild animals are the main causes of infection [24]. For example, a sero-prevalence study conducted by Schoonman and colleagues [93] revealed that, among livestock farmers, veterinary/meat inspectors, and abattoir workers, the prevalence of Leptospira spp. infection was 19.4, 18.1, and 17.1%, respectively. Additionally, subjects who reported milking cows had a significantly higher likelihood of testing positive for the infection [47].

### 1.6 Pathogenesis and clinical sign

Despite decades of study on leptospirosis, understanding mechanisms of leptospirosis pathogenesis is limited [94]. For example, the cause of the severity or mildness of an infection, whether caused by the pathogen directly or by host immune responses that are genetically defined, is yet unclear [95, 96]. However, the emergence of genetic manipulation instruments and the release of genome sequences from saprophytic and pathogenic *Leptospira* have recently provided new information about the biology and pathophysiology of this significant agent [97]. *Leptospira's* motility and capacity to float through viscous liquids are two virulence mechanisms [98].

The preliminary contagion and the spread of organisms from the site of entrance to end-organ damage locations, including the lung, liver, kidney, eye, and brain, are likely critical processes involving motility. Among the *Leptospira* genome sequence, there are 4768 predicted genes, at least 50 of which are related to motility [99]. After entering a new host, Leptospira reach the kidneys through the hematogenous route, passing glomeruli or peritubular capillaries. They overrun the brush boundary of the proximal tubule, which is where urine is excreted, when they penetrate the renal tubular lumen [24]. Pathogenic *Leptospira* can inflict local or systemic harm if it enters the bloodstream. Variations in severity are related to patient age, health state, inoculum size, and serovars [17].

*Leptospira* incubation lasts approximately 10 days, and the illness manifests 5–14 days following the bacteria's introduction [100]. A wide range of events may be involved in the acute and chronic infection processes of humans and reservoir hosts, according to the widely varying clinical presentations of *Leptospira* infection [101]. *Leptospira* spread quickly and multiplied in the bloodstream during the early phase (3–7 days), after which they went to the kidney, liver, spleen, and other organs [17]. Pathogenic *Leptospira* infection in hosts results in a wide range of clinical symptoms, including subclinical infection, undifferentiated febrile illness, jaundice, renal failure, and possibly fatal pulmonary hemorrhage [24]. On the other hand, conjunctival suffusion is thought to be a distinctive clinical feature of the early stages of leptospirosis [17]. The proximal renal tubules of carrier animals continue to be colonized, which sustains leptospirosis. An animal that is afflicted may not show any symptoms and may secrete infectious organisms in its urine for the whole of its life [102, 103].

The clinical course of leptospirosis can be split into two phases: the acute phase, marked by flu-like symptoms that are milder, and the chronic phase, which has more severe clinical indications, including renal failure and pulmonary diffuse hemorrhage [100, 104]. In certain instances, the infection may not cause any symptoms. The majority of individuals infected with the pathogen either exhibit few symptoms, such as fever, headache, myalgia, nausea, vomiting, malaise, and conjunctival hyperaemia, or none at all [105]. However, as one of the many potential causes of acute fevers in medical settings, patients experience around 7 days of fever, photophobia, headache, arthralgia, chills, diaphoresis, asthenia, cough, nausea, vomiting, and muscle discomfort that can all manifest during the acute phase of the leptospirosis stage. These symptoms are similar to those of other illnesses like dengue and influenza, which cause misdiagnosis and underreporting of leptospirosis cases worldwide. If treatment is not received, serious forms may develop, resulting in serious pulmonary hemorrhage, liver damage, and renal damage [104]. Leptospirosis motility is likely significant for both the initial infection and the spread of the organisms from the point of entry to end-organ damage sites, including the liver, kidney, brain, lung, and eye [106]. It has been demonstrated that virulent Leptospira strains have a chemotaxis toward haemoglobin [107]. Leptospira also has a variety of putative virulence factors, including phospholipase, sphingomyelinase, and haemolytic activities, which are consistent with their projected capacity to move through host tissues [108].

According to estimates [17], between 10 and 15% of tainted individual's exhibit major clinical signs, which are more frequent in the late phase (after the first week of infection). However, occasionally, even in the early stages, very severe signs are noted. Weil's syndrome is the term for the characteristic symptom of the severe form of leptospirosis. Chronic interstitial nephritis, pancreatitis, anemia, splenomegaly, jaundice, renal failure, myocarditis, and pulmonary hemorrhage are among the conditions that these patients may experience [109]. Interstitial nephritis is the key pathological change detected in most cases of patients [110]. According to Ferrer et al. [111], experimental mouse models of Leptospira outer membrane proteins (OMPs) cause local inflammation by stimulating renal cells to produce pro-inflammatory cytokines. Furthermore, one of the most alarming side effects of leptospirosis is severe pulmonary hemorrhage syndrome (SPHS), which is characterized by cough, dyspnoea, haemoptysis, and dense pulmonary hemorrhage as its primary symptoms. About half of the patients with SPHS pass away as a result of these problems [112]. Leptospirosis in cattle is a reproductive disease that is widely known. The main features of this illness include poor offspring's birth and abortions occurring at any time during a pregnancy. There have been reports of nursing cows' milk tainted with blood. Conversely, the most typical sign is a silent, on-going infection associated with recurrent estruses and the loss of embryos. Due to their capacity to colonize and bear in the genital tracts of afflicted cows and bulls, serovars Hardjo and Guaricura are more closely linked with the chronic reproductive type of disease. It is becoming well-acknowledged that the main sign of leptospirosis affecting cattle is this genital manifestation of the disease [113–115].

### 1.7 Diagnosis and treatment

In endemic areas, leptospirosis infection has a very broad spectrum, which might complicate diagnosis and is frequently asymptomatic or symptomless [68]. As a result, the disease is not as well-known as it could be, and diagnosis is not provided quickly enough or with sufficient accessibility. Leptospirosis is diagnosed depending on the temporal stage of the disease and the sample's availability. Laboratory techniques such as microscope evaluation, culture, molecular approaches, serology, and animal inoculation are employed to detect Leptospira [116]. Bacterial culture is the gold-standard diagnostic technique, but due to financial and technological constraints, it is not widely used. Nonetheless, epidemiologic investigations of animal strains infecting a particular area need the isolation and genetic characterization of isolates [117, 118]. Thus, an immunoglobulin M (IgM) serological screening approach on a herd level with usually enzyme-linked immune-sorbent assay (ELISA), or MAT may identify suspect herds, but culturing, immunofluorescence, or, preferentially, polymerase chain reaction (PCR) assay is required for a definitive diagnosis in acute phase leptospirosis [91]. Although urine is primary and has been extensively used as a sample, recent studies have demonstrated the usefulness of cervicovaginal mucus for the diagnosis of genital infection. The mortality rate can be significantly decreased by starting antimicrobial medication early and receiving supportive therapy right away [91]. Thus, early detection is essential to lower the severity and rate of disease progression since fast treatment reduces mortality. *Leptospira* has not shown to have acquired antibiotic resistance, despite slight variations in antibiotic susceptibility among isolates so far [119]. Amoxicillin (or penicillin), tetracycline, or ceftriaxone, doxycycline, streptomycin, and erythromycin are used to treat leptospirosis [60, 72, 120, 121]. Early antibiotic therapy has been shown to reduce the risk of severe disease in both the individual and the population, as well as to reduce the severity and fatality rates and save costs for the healthcare system [60]. Prompt presumptive antibiotic treatment is also advised by the WHO, citing the empirical advantage of starting it prior to the fifth day of illness [72]. The requirements for supportive care, including intensive care, will vary depending on the severity of leptospirosis. The majority of cases of severe leptospirosis will necessitate some kind of supportive care, including intensive care. Hemodynamic instability is often evident in the presentation, necessitating the use of vasoactive medications and intravenous therapy, which will also be employed to administer medication and balance correction. Corticosteroids are occasionally utilized because patients often appear with a septic shock, and some degree of excessive inflammatory response is reported [122].

### 1.8 Prevention and control measures

The adoption of vaccination is highly recommended since it is the least expensive control strategy and is crucial for the management of leptospirosis [123]. Vaccinating both humans and animals helps prevent leptospirosis. Its effectiveness varies, and research has shown that commercial vaccinations are ineffective at preventing kidney colonization [124]. The effectiveness of the control methods differs depending on the strain of infection. Despite cross-reactivity, some vaccinations may provide protection against serovars, unlike those used to manufacture them [125]. However, their potential serotype dependence and inability to elicit an immunological response against other serovars were noted by Sonrier et al. [126]. Furthermore, the creation of vaccines is hampered by the genetic and phenotypic variety of infective Leptospira [127].

According to Pereira et al. [18], Cuba is the only country in Latin America where people are routinely immunized against Leptospira. Inactivated vaccinations are typically given to domestic animals. It is also used on populations of susceptible people in numerous other nations. But they are not employed everywhere because of unpleasant effects on humans, a lack of cross-protection, and a short duration of immunity [127]. Mostly, the leptospirosis vaccine schedule typically calls for six-month intervals, with the vaccination window falling preferably before the breeding service season, which starts in the early spring when environmental contamination is at its peak [105]. Sejroe strain infections, on the other hand, are impossible to completely eradicate; instead, on-going vigilance and a programme centred on lowering reproductive issues and the ensuing financial risks are necessary [128]. Incidental infections can be managed more readily. The other leptospirosis control techniques are designed to reduce the incidence of clinical disease by implementing integrated actions in many parts of the transmission chain [62, 129].

## 2. Challenging factors for leptospirosis control and prevention

It is anticipated that leptospirosis risk will increase in the future [24, 41, 56, 60]. Different literature classifications of leptospirosis risk variables include behavioral, climatic, animal exposure, physical, environmental, and socioeconomic threats. Particularly in tropical and subtropical nations, climate change is expected to exacerbate extreme weather conditions and flooding, which could lead to an increase in the frequency of outbreaks [89]. According to research by Dhewantara [130], one of the elements supporting the emergence of leptospirosis' global dissemination is climate. Some other challenging factors for leptospirosis control and prevention are the biological characteristics of the etiological agent, the reservoir host, and differential diagnosis.

### 2.1 Biological characteristics of *Leptospira* as a major challenge

The mechanism of cell aggregation as an endurance technique for *leptospires* was detected by Trueba et al. [131]. Through the mechanism of cell aggregation, the bacteria can accumulate nutrients and enzymes from the lysing cells, which can help them, survive hard conditions, such as low temperatures or nutrient-poor environments [86, 132]. *Leptospira* have a double-membrane that is similar to that of gram-negative bacteria due to the presence of lipopolysaccharide (LPS), and their cytoplasmic membrane closely resembles the envelope architecture of Gram-positive bacteria due to the tight interaction of the peptidoglycan cell wall [80]. On the other hand, *Leptospira* shares a thin peptidoglycan cell wall with other gram-negative bacteria. As noted by Miyamoto-Shinohara et al. [133], the thickness of the cell walls surely affects the survival rate of bacteria. LPS is another characteristic of *Leptospira* species. By keeping water within the cells, LPS aids in the survival of bacteria in the desiccant environment [134, 135]. For instance, to avoid ice forming inside the cells and moisture build-up during the storage of dried culture, all water must be completely removed from the cell. Accordingly, LPS prevents water molecules from being extracted from inside the cells during the freeze-drying procedure [133]. Because of their meticulous nature, *Leptospira* spp. are difficult to isolate, propagate, and store [136].

### 2.2 Diagnostic challenges

Leptospirosis in humans is frequently misdiagnosed and disregarded as the source of fever or generalized sickness. Since it can mimic many other infectious diseases, including dengue, malaria, hepatitis, pneumonia, and meningitis, misdiagnosis is widespread due to its varied symptoms and non-specific presentations. Other illnesses such as haemorrhagic fevers, Q fever, scrub typhus, yellow fever, and rickettsia infections may be included in the differential diagnosis, depending on the patient's geographic location [137–140]. Following natural catastrophes, infectious disease epidemics are frequent [138–140]. Concurrent outbreaks of various pathogens can pose diagnostic hurdles, especially if laboratory facilities are limited. The clinical ramifications of misdiagnosis or delayed diagnosis are noteworthy, as the reduction of morbidity and death associated with early treatment of certain infections like leptospirosis, malaria, and meningitis is imperative. Furthermore, misdiagnosis can cause significant delays in the application of therapies specific to a given disease, giving epidemics the opportunity to "escape." Following flooding, food- and water-borne diseases are frequent, and some of them might be challenging to distinguish clinically from leptospirosis [89].

Diagnosis is based on laboratory results. However, the clinical appearance of leptospirosis is unclear, and the laboratory diagnosis of leptospirosis is often challenging. The polymerase chain reaction (PCR), which is infrequent in high-endemic areas, is the only sensitive and specific diagnostic that is accurate during the acute phase of the disease [141, 142]. Micro-agglutination testing (MAT), a serological reference approach, is only accessible in Ref. laboratories. Leptospirosis may be mistaken for malaria, influenza, dengue, scrub or murine typhus, spotted fevers, and a variety of other potential viral, parasitic, or bacterial diseases in tropical environments where there are many potential causes for acute fevers [60]. Among hundreds of cases of influenza, dengue, or yellow fever, leptospirosis may be the cause of a few severe cases of fever. As leptospirosis lacks a pathognomonic sign or symptom, samples should be sent to a laboratory for biological study, and the epidemiological perspective should be used to raise suspicions [60, 118]. It has been possible to detect motile, helical-shaped Leptospira bacilli directly in blood or urine; however, this method has been shown to have low sensitivity and is difficult to determine specificity due to the bacteria's small diameter [44]. Therefore, caution should be exercised while using this technique. These days, genes specific to pathogenic Leptospira are targeted by nucleic acid amplification techniques using a matrix of blood, urine, or cerebrospinal fluid. In order to effectively control this potentially fatal zoonosis, an efficient surveillance system that tracks disease trends must be developed.

Our understanding of the disease will grow with the use of these data, but improving veterinary and public health surveillance and reporting will require more laboratory space and the resolution of diagnostic hurdles by scientists. To full-fill monitoring needs and prevent and reduce leptospirosis epidemics, a collaborative, multidisciplinary One Health approach is required [8]. Early detection of animal sickness could be used to detect environmental pollution and initiate preventive actions to lower the risk of leptospirosis infection in humans [54]. Reducing the opportunities for wild animals to engage with domestic and companion animals, utilizing potable water, immunizing livestock and companion animals, and donning water-proof protection gear and footwear when near potentially polluted water are all possible ways to prevent infection. To determine whether an activity poses a risk of infection for their pet, to learn about infection signs, and to take protective measures in case their pet becomes ill, pet owners should consult with their veterinarian [136].

Organizations dedicated to public and animal health should think about doing educational outreach in places that have recently seen infections in humans or animals, that occasionally flood, and that include recreational water areas or events that could potentially spread the disease. As was previously noted, leptospirosis is primarily found in tropical and subtropical areas, where there are several potential causes of acute fevers. In situations where a hospital has limited space, patient triage may be necessary, particularly in critical care units. This is especially prevalent in two scenarios: either in conjunction with an on-going (usually viral) outbreak or during large-scale leptospirosis outbreaks that follow cyclones [143]. In this specific case of concurrent dengue fevers, leptospirosis should receive extra attention due to its challenging differential diagnosis and the possibility of co-infections [60]. Leptospirosis should receive extra attention in this specific setting of concurrent dengue fevers due to its challenging differential diagnosis, especially given the possibility of co-infections [60]. To find biological indicators with a prognostic value for severity, several clinical investigations involving leptospirosis patients have been carried out. Validated prognostic markers for leptospirosis are still up for debate, despite certain cohorts demonstrating significant differences.

## 2.3 Animal reservoirs

Animals are traditionally categorized as susceptible or maintenance hosts. Susceptible hosts typically experience a variable-severity illness after infection, which may even be fatal. Within a few weeks, the recuperating hosts will completely eradicate the *Leptospira* from their bodies. After contagion, vulnerable hosts usually show a course of sickness with a range of severity; sometimes, it may be fatal. The illness varies in severity. Within weeks, the recovered hosts will totally clear their bodies from the *Leptospira*. With rare exceptions, the maintenance hosts will likely neither acquire nor exhibit any mild clinical symptoms, and they will also likely rid their bodies of *Leptospira* with the exception of the kidneys, where they will likely continue to grow and persist in the proximal renal tubules for several months.

Preservation host populations are called "reservoirs" for leptospirosis. Among the various Leptospira, different serovars with frequent host associations cattle with the serogroup Hardjo; dogs and canines with the serogroup Canicola; mice with the serogroup Ballum; rats with the serogroup Icterohaemorrhagiae are common. Recent research on the *Rattus norvegicus* model leptospirosis reservoir has also revealed evidence of Leptospira in the breast tissue and milk of females with long-term infections. Even though it has already been suggested that leptospirosis can spread through breastfeeding [144]. Conditions for the *Leptospira* strains implicated in individuals' cases are determined by the animal reservoir in a particular setting. Many mammals in rural areas, such as domestic animals like pigs, cattle, dogs, sheep, and goats, preserve and disperse a variety of *Leptospira* serovars. Conversely, in typical urban and peri-urban leptospirosis, the main reservoir is rats (mainly *Rattus norvegicus*) bearing the serogroup Icterohaemorrhagiae. As "vectors," domestic dogs may, for example, be seen wandering in areas where they come into contact with soil, water, or wild or feral creatures. Within days to weeks, they may become infected and start to excrete *leptospires* in their urine, even in close proximity to their owners.

### 2.4 Environmental drivers of leptospirosis

Notwithstanding the prevalence and range of leptospirosis, there are still a great deal of unanswered questions regarding the dynamics of the disease's transmission and the catalysts for epidemics. However, several environmental risk factors for illness or outbreaks have been found by epidemiological investigations, and these factors vary depending on the ecological area. Urbanization and climate change are probably going to have an impact on a lot of these variables.

#### *2.4.1 The environmental reservoir*

*Leptospira* are released into the environment along with the urine of infected animals. It is believed that pathogen *Leptospira* exist in the environment but does not proliferate [44, 145]. Early research has identified a few physiological characteristics of Leptospira that influence their ability to survive in the environment [15, 146, 147]. It is commonly known that pathogen *Leptospira* strain can thrive in neutral to slightly alkaline soils and freshwater environments, such as muck, swamps, streams, lakes, and rivers. Both humans and animals can get infected if they come into contact with that kind of muck or stagnant water. When there is a lot of rain, the surface soils become washed away, exposing humans to pathogenic *leptospiras* in freshwater bodies like floods. Large-scale outbreaks of leptospirosis typically occur after cyclones or other periods of intense rainfall and flooding [148–150]. Notably, the continuous changes in climate will likely affect leptospirosis prevalence and distribution in some parts of the world by altering patterns of rainfall [89]. Infection of humans directly from the animal reservoir occurs primarily in work environments. Particularly workers who deal with rodent management, farmers, hunters, butchers, and veterinarians, among other occupations, may come into direct contact with the contaminated urine. Still, data indicates that the majority of human infections are caused by indirect exposure to surface waters, soils, and mud. Occupational exposures among freshwater fishermen, sewage workers, miners, soldiers, farmers, and farmers operating in irrigated rice or taro fields, sugarcane, or banana farms are also accounted for by indirect environmental exposure. Mucosae and skin wounds are entrance points for Leptospira into the body.

### *2.4.2 Habitats and animal interactions*

Habitats that allow for interspecies interactions may promote transmission. Conditions that facilitate interactions between wildlife and farm animals may be produced by agricultural operations. According to New et al., deer frequently consume food that has been left out for cattle [151]. According to Jenkins and associates, swine farms frequently draw stray canines [152]. Among the 30 environments examined, Brown and Gorman conjectured that *Leptospira* was nearly exclusively found in areas with substantial populations of house mice (*Mus musculus*) [153]. The increased host susceptibility density could support the persistence of the illness. Domestic cat ownership may benefit human occupants by lowering rodent numbers in residential settings, according to one study [154].

### *2.4.3 Host and management risk factors*

Animals of any age can get a *Leptospira* infection. While leptospirosis affects almost all mammalian species, it is most frequently found in pigs, dogs, horses, sheep, and goats [62, 155]. Risk factors for infection include the introduction of infected animals into herds, sharing grazing areas with infected animals, contact with infected water sources like rivers, flood waters, or drainage water, and reproduction with infected male animals by natural insemination [156]. Leptospirosis has been linked to animal exposure, especially to rats, rodents, household pets, and livestock, as would be predicted given its route of transmission [24]. The animals that are linked to the spread of disease carry different serovars in different geographical areas; exposure risk is influenced by a number of factors, including living conditions, agricultural activities, local wildlife, and cultural norms.

### *2.4.4 Climatic change and seasonal factors*

The transmission of leptospirosis is frequently seasonal, peaking in warm regions during the rainy seasons and in temperate climates during the summer [12]. Seasonal trends were also identified in veterinary diagnostic studies [157]. Flooding and excessive rains contribute to increasing the threat of leptospirosis. Across the globe, in a variety of geographical locations, there have been numerous reports of leptospirosis outbreaks following extreme weather events [89]. Over the last 50 years, all the scenarios involving greenhouse emissions have resulted in an extraordinary rise in the global surface temperature, and this trend has persisted until the next mid-century [158]. These shifting circumstances can affect hosts, pathogens, vectors, and the spread of disease [159, 160]. In particular, infections can only grow in a limited range of temperatures; they cannot grow at higher or lower temperatures. Higher temperatures hasten the pathogens maturation, which in turn affects the pathogens proliferation and extrinsic incubation duration within a vector.

Climate change poses a number of health risks, and these can be classified as biomedical (i.e. immune-compromised and malnourished), demographic (i.e. age and sex), geographic (i.e. land use and flood zones), socioeconomic (i.e. poverty and education), or socio-political (i.e. civil unrest and political instability) [161].

A weather or climate event can initiate cascading risk pathways, which are a series of secondary, causally connected events, in addition to these effects on infections, vectors, hosts, and disease transmission [160]. In high-risk counties in China, rainfall and land surface temperature were found to be substantially correlated with the notification of human leptospirosis between 2006 and 2016 [162]. Leptospirosis incidences can be brought on by infected drinking water or floodwater [160, 163], clearly demonstrating how population exposure, societal vulnerability, and climatic hazards interact to create cascading risk pathways for leptospirosis.

### *2.4.5 Rainfall and flooding*

The most frequent natural disaster, flooding, can raise the risk of infectious diseases like leptospirosis by interfering with public health services and infrastructure, destroying water and sanitation systems, uprooting populations, causing property damage, and increasing the exposure of the environment to pathogens [89, 164]. The bulk of the world's most vulnerable populations reside in Asia, the Pacific Islands, Africa, and the Middle East, where 70% of these disasters happen [165]. Rainfall and flooding contribute to the proliferation of rats, the dispersal of food, trash, and debris, the disturbance of sewage and waste management systems, and the promotion of vegetation growth that raises the availability of food [89, 166].

In addition, floods enhance animal-human interaction when people are uprooted from their houses and animals are forced from their customary territories [167].

It is true that intricate socio-ecological processes play a role in leptospirosis transmission. In tropical and subtropical areas, the majority of leptospirosis epidemics happen during the monsoon or humid, wet seasons. These conditions create an ideal environment for Leptospira transmission by bringing in more water, moist soil, and flooding [60, 162]. Certain pathogen *Leptospira* strains have been shown to survive for days or even months in such favorable environmental circumstances [168, 169]. Additionally, in rural areas, the rainy season coincides with a period of high farming activity (e.g. planting or irrigating paddy fields, harvesting and herding). Simultaneously, the number of mice in the ecosystem grows [170] due to increased rainfall and increased productivity of the land and plants, which in turn provide more food and an environment conducive to rodent breeding. Leptospira shedding into the environment and the risk of transmission among reservoir animals may both increase with a rise in the rodent population. In addition, inadequate waste management, unsanitary practices, and risky behaviors (such as going barefoot) brought on by low socioeconomic circumstances raise the risk of infection in humans [101]. As a consequence, a non-specific clinical presentation may occur, and leptospirosis is frequently misdiagnosed, particularly in areas where scrutiny and analytical abilities are inadequate. Assessing the long-term effects of leptospirosis after floods requires continuous monitoring during the post-disaster phase [166].

## 3. Temperature

Higher temperatures and humid settings allow *Leptospira* to live longer, and seasonal variations in leptospirosis incidence are well documented [171, 172]. Increased temperatures can also lessen the availability of surface water due to evaporation while also promoting animal and human water-based activities (such as drinking, swimming, and bathing). This increases the amount of time that people, livestock, pets, and wildlife spend together by sharing the limited surface water sources [89]. The interaction of rainfall, surface runoff, sea level, catchment size, and local terrain is triggered by the consequences of flooding. Land use, urbanization, deforestation, irrigation, agricultural techniques, dams, and water management are some of the ways that these components might change in turn [173]. Increasing sea levels, rising land and sea surface temperatures, a rise in the frequency of hazardous weather events, stronger tropical cyclones, and greater storm surges are additional factors that will increase the danger of floods due to global climate variance [89]. A correlation has also been shown in certain studies [174, 175] between the incidence of leptospirosis and temperature and humidity.

Nowadays, wide ranges of physical environment data at different spatial and temporal scales are delivered by remote-sensing (RS) technologies, which can aid in the understanding of disease epidemiology [176, 177]. In Kampong Cham province, Cambodia, the modified normalized difference water index (MNDWI), which is produced from the Moderate Resolution Imaging Spectroradiometer (MODIS), has been used as a flood indicator and to assess the risk of leptospirosis [178].

## 4. Conclusions and future prospect

The complexity and variability of leptospirosis epidemiology vary significantly in various environmental settings. The global burden of leptospirosis disease will most likely increase as a result of climate change, flooding, population growth, and urbanization. Understanding disease patterns from an ecological perspective, like the local, regional, and global levels, hence requires. Comprehending the environmental factors that contribute to leptospirosis infection is also crucial for enhancing community awareness and hazard mitigation, as well as developing local capacity to get ready for the rising leptospirosis risk brought on by climate change. More comprehensive surveillance and research are required to better understand leptospirosis transmission patterns and how they may be impacted by climatic events, environmental variables, animal reservoirs, and demographic and socioeconomic trends in humans. Additionally, establishing instruments to determine the causes of outbreaks, create early warning systems, project the frequency and severity of outbreaks, and pinpoint regions most susceptible to a high disease burden can also be important. If such data is available, emergency and disaster response stakeholders could use it to manage risk factors at disaster sites, improve the timeliness of emergency responses, assess the ability of response teams to meet future demands, provide an evidence base for disaster management and public health resource allocation planning and allocation, and concentrate interventions on reducing the risk of infection and the overall disease burden from leptospirosis.

Apart from the existing burden, data indicates that leptospirosis incidence may rise due to climate change, particularly but not exclusively in temperate zones. There is currently no information available on the true cost of this resurgent zoonosis in a number of developing nations with lifestyle, climate, and ecological factors that make leptospirosis likely. It is also likely to be overlooked in many of these nations due to its ambiguous clinical presentation and complex biological diagnosis. Wearing protective gear, such as gloves, eye protection, and face masks, is advised while handling infected animals or working in contaminated regions. Ethiopia and other African nations lack adequate information about leptospirosis. While leptospirosis is easily treatable with most antibiotics, it is also simple to diagnose using microscopy, culture, and serological methods. Antibiotics are an efficient way to treat leptospirosis, and vaccination against the disease can reduce the severity of the illness in domestic animals. The current leptospirosis policies emphasize early case detection and timely treatment. Undoubtedly, this limits mortality and problems in their early stages, but when combined with an integrated rodent management strategy, this can be strengthened even further. It may be helpful to identify risk zones, such as popular tourist destinations, fields, rivers, and streams, and to raise awareness of these areas by highlighting vulnerable populations.

In the upcoming contention for strategy and programme, the integration of rodent control is still lacking in underdeveloped nations, despite an abundance of measures aimed at preventing tropical illnesses that are increasingly being overlooked. Therefore, research is also required to comprehend the potential reservoir due to the variety of causes. The creation of computerized databases, the merging of government and commercial hospitals with academic research groups, and the use of geographic information systems can all be useful in figuring out how the disease is distributed throughout the globe. Frontline healthcare providers may be investigated as a means of educating the public and equipping them with the fundamental knowledge and abilities needed to identify risk factors and report them on time.

## Acknowledgements

I want to acknowledge the IntechOpen for giving me a chance to contribute to this public, livestock, environmental and socioeconomically critical disease for the scientific community.

## Conflict of interest

The author declares no conflict of interest.

## Author details

Aregitu Mekuriaw
Research and Development Directorate, Molecular Biology Laboratory, National Veterinary Institute, Bishoftu, Ethiopia

*Address all correspondence to: aregitumekuriaw@gmail.com

IntechOpen

© 2025 The Author(s). Licensee IntechOpen. This chapter is distributed under the terms of the Creative Commons Attribution License (http://creativecommons.org/licenses/by/3.0), which permits unrestricted use, distribution, and reproduction in any medium, provided the original work is properly cited.

## References


[1] Coppola F, Cilia G, Bertelloni F, Casini L, D'Addio E, Fratini F, et al. Crested porcupine (*Hystrix cristata* L.): A new potential host for pathogenic Leptospira among semi-fossorial mammals. Comparative Immunology, Microbiology and Infectious Diseases. 2020;**70**:101472

[2] Agampodi S, Gunarathna S, Lee JS, Excler JL. Global, regional, and country-level cost of leptospirosis due to loss of productivity in humans. PLoS Neglected Tropical Diseases. 2023;**17**(8):e0011291

[3] Burns DS, Clay KA, Bailey MS. Leptospirosis in a British soldier after travel to Borneo. BMJ Military Health. 2016;**162**(6):473-475

[4] Lacerda HG, Monteiro GR, Oliveira CC, Suassuna FB, Queiroz JW, Barbosa JD, et al. Leptospirosis in a subsistence farming community in Brazil. Transactions of the Royal Society of Tropical Medicine and Hygiene. 2008;**102**(12):1233-1238

[5] Mgode GF, Japhary MM, Mhamphi GG, Kiwelu I, Athaide I, Machang'u RS. Leptospirosis in sugarcane plantation and fishing communities in Kagera northwestern Tanzania. PLoS Neglected Tropical Diseases. 2019;**13**(5):e0007225

[6] Yilmaz O, Celebi G, Atabek E, Kokturk F, Atakent SD, Piskin N, et al. Seroprevalence of leptospirosis in coal miners in Zonguldak region/Zonguldak Bolgesinde Komur Madeni Iscilerinde Leptospiroz Seroprevalansi. Klimik Journal. 2019;**32**(2):174-178

[7] Khaki P, Khodaverdi DE, Moradi BS, Yahaghi H, Dejbord M, Soltani MN. An assay for determining the epidemy of dominant leptospirosis producing serovars among personnel of an industrial slaughter house. Journal of Microbial World. 2012;**4**(3&4):101-108

[8] Hernández-Rodríguez P, Trujillo-Rojas B. One health: A comprehensive approach to improve prevention and control strategies in leptospirosis. Revista de Ciências Agroveterinárias. 2022;**21**(1):71-78

[9] Garba B, Bahaman AR, Zakaria Z, Bejo SK, Mutalib AR, Bande F, et al. Antigenic potential of a recombinant polyvalent DNA vaccine against pathogenic leptospiral infection. Microbial Pathogenesis. 2018;**124**:136-144

[10] Biscola NP, Fornazari F, Saad E, Richini-Pereira VB, Campagner MV, Langoni H, et al. Serological investigation and PCR in detection of pathogenic leptospires in snakes. Pesquisa Veterinária Brasileira. 2011;**31**:806-811

[11] Jobbins SE, Alexander KA. Evidence of *Leptospira* sp. infection among a diversity of African wildlife species: Beyond the usual suspects. Transactions of the Royal Society of Tropical Medicine and Hygiene. 2015;**109**(5):349-351

[12] Levett PN, Morey RE, Galloway RL, Turner DE, Steigerwalt AG, Mayer LW. Detection of pathogenic Leptospires by real-time quantitative PCR. Journal of Medical Microbiology. 2005;**54**(1):45-49

[13] Thibeaux R, Girault D, Bierque E, Soupé-Gilbert ME, Rettinger A, Douyère A, et al. Biodiversity of environmental Leptospira: Improving identification and revisiting the diagnosis. Frontiers in Microbiology. 2018;**9**:816

[14] Karpagam KB, Ganesh B. Leptospirosis: A neglected tropical zoonotic infection of public health importance—An updated review. European Journal of Clinical Microbiology & Infectious Diseases. 2020;**39**(5):835-846

[15] Warnasekara J, Agampodi S, Abeynayake R, R. Time series models for prediction of leptospirosis in different climate zones in Sri Lanka. PLoS One. 2021;**16**(5):e0248032

[16] Bashiru G, Bahaman AR. Advances & challenges in leptospiral vaccine development. Indian Journal of Medical Research. 2018;**147**(1):15-22

[17] Azevedo IR, Amamura TA, Isaac L. Human leptospirosis: In search for a better vaccine. Scandinavian Journal of Immunology. 2023;**98**(5):e13316

[18] Prager KC, Greig DJ, Alt DP, Galloway RL, Hornsby RL, Palmer LJ, et al. Asymptomatic and chronic carriage of Leptospira interrogans serovar Pomona in California Sea lions (*Zalophus californianus*). Veterinary Microbiology. 2013;**164**(1-2):177-183

[19] Dietrich M, Wilkinson DA, Benlali A, Lagadec E, Ramasindrazana B, Dellagi K, et al. L eptospira and paramyxovirus infection dynamics in a bat maternity enlightens pathogen maintenance in wildlife. Environmental Microbiology. 2015;**17**(11):4280-4289

[20] da Rocha Albuquerque M, dos Reis TA, de Souza Rocha K, da Silva Brito J, de Sena Mesquita GS, da Silva AF, et al. Evidence of exposure to *Leptospira* spp. in dogs at a zoonosis control center in Brazil. Revista Acadêmica Ciência Animal. 2020;**18**:1-6

[21] Jansen A, Stark K, Schneider T, Schöneberg I. Sex differences in clinical leptospirosis in Germany: 1997-2005. Clinical Infectious Diseases. 2007;**44**(9):e69-e72

[22] Dreyfus A, Dyal JW, Pearson R, Kankya C, Kajura C, Alinaitwe L, et al. Leptospira seroprevalence and risk factors in health Centre patients in Hoima District, Western Uganda. PLoS Neglected Tropical Diseases. 2016;**10**(8):e0004858

[23] Lagadec E, Gomard Y, Le Minter G, Cordonin C, Cardinale E, Ramasindrazana B, et al. Identification of Tenrececaudatus, a wild mammal introduced to Mayotte Island, as a reservoir of the newly identified human pathogenic Leptospira mayottensis. PLoS Neglected Tropical Diseases. 2016;**10**(8):e0004933

[24] Haake DA, Levett PN. Leptospirosis in humans. Leptospira and Leptospirosis. 2015;**387**:65-97

[25] Ellis WA. Animal leptospirosis. Leptospira and Leptospirosis. 2015;**387**:99-137

[26] Guerrant RL, Walker DH, Weller PF. Tropical Infectious Diseases. Pathology. Elsevier Inc.; 2006

[27] Taylor C, Brodbelt DC, Dobson B, Catchpole B, O'neill DG, Stevens KB. Spatio-temporal distribution and agroecological factors associated with canine leptospirosis in Great Britain. Preventive Veterinary Medicine. 2021;**193**:105407

[28] Nava A, Shimabukuro JS, Chmura AA, Luz SLB. The impact of global environmental changes on infectious disease emergence with a focus on risks for Brazil. ILAR Journal. 2017;**58**(3):393-400

[29] Reis RB, Ribeiro GS, Felzemburgh RD, Santana FS,

Mohr S, Melendez AX, et al. Impact of environment and social gradient on Leptospira infection in urban slums. PLoS Neglected Tropical Diseases. 2008;**2**(4):e228

[30] Dung LP, Hai PT, Hoa LM, Mai TNP, Hanh NTM, Than PD, et al. A case–control study of agricultural and behavioral factors associated with leptospirosis in Vietnam. BMC Infectious Diseases. 2022;**22**(1):583

[31] Bouscaren N, de Coignac CB, Lastère S, Musso D, Teissier Y, Formont J, et al. Leptospirosis in French Polynesia: 11 years of surveillance data, 2007-2017. New Microbes and New Infections. 2019;**29**:100518

[32] Durski KN, Jancloes M, Chowdhary T, Bertherat E. A global, multi-disciplinary, multi-sectorial initiative to combat leptospirosis: Global leptospirosis environmental action network (GLEAN). International Journal of Environmental Research and Public Health. 2014;**11**(6):6000-6008

[33] Ricaldi JN, Vinetz JM. Leptospirosis in the tropics and in travelers. Current Infectious Disease Reports. 2006;**8**:51-58

[34] Bandara M, Ananda M, Wickramage K, Berger E, Agampodi S. Globalization of leptospirosis through travel and migration. Globalization and Health. 2014;**10**:1-9

[35] De Vries SG, Visser BJ, Stoney RJ, Wagenaar JF, Bottieau E, Chen LH, et al. Leptospirosis among returned travelers: A GeoSentinel site survey and multicenter analysis—1997-2016. The American Journal of Tropical Medicine and Hygiene. 2018;**99**(1):127

[36] Schönfeld A, Jensen B, Orth HM, Tappe D, Feldt T, Häussinger D. Severe pulmonary haemorrhage syndrome in leptospirosis in a returning traveller. Infection. 2019;**47**:125-128

[37] Walker MD. Leptospirosis: The possible risk to those participating in water-based sports and activities. British Journal of General Practice. 2018;**68**(673):394-395

[38] Soo ZMP, Khan NA, Siddiqui R. Leptospirosis: Increasing importance in developing countries. Acta Tropica. 2020;**201**:105183

[39] Morand S, Jittapalapong S, Kosoy M. Rodents as hosts of infectious diseases: Biological and ecological characteristics. Vector Borne and Zoonotic Diseases. 2015;**15**(1):1

[40] De Vries SG, Visser BJ, Nagel IM, Goris MG, Hartskeerl RA, Grobusch MP. Leptospirosis in sub-Saharan Africa: A systematic review. International Journal of Infectious Diseases. 2014;**28**:47-64

[41] Costa F, Hagan JE, Calcagno J, Kane M, Torgerson P, Martinez-Silveira MS, et al. Global morbidity and mortality of leptospirosis: A systematic review. PLoS Neglected Tropical Diseases. 2015;**9**(9):e0003898

[42] Vincent AT, Schiettekatte O, Goarant C, Neela VK, Bernet E, Thibeaux R, et al. Revisiting the taxonomy and evolution of pathogenicity of the genus *Leptospira* through the prism of genomics. PLoS Neglected Tropical Diseases. 2019;**13**(5):e0007270

[43] World Health Organization. Report of the second meeting of the leptospirosis burden epidemiology reference group. 2011

[44] Samrot AV, Sean TC, Bhavya KS, Sahithya CS, Chan-Drasekaran S, Palanisamy R, et al. Leptospiral infection, pathogenesis and its

diagnosis—A review. Pathogens. 2021;**10**(2):145

[45] Torgerson PR, Hagan JE, Costa F, Calcagno J, Kane M, Martinez-Silveira MS, et al. Global burden of leptospirosis: Estimated in terms of disability adjusted life years. PLoS Neglected Tropical Diseases. 2015;**9**(10):e0004122

[46] Pal M, Hadush A. Leptospirosis: An infectious emerging waterborne zoonosis of global significance. Air & Water Borne Diseases. 2017;**6**(1):1-4

[47] Mteshana ZC. Epidemiology of Leptospirosis in Groote Schuur Hospital. 2021

[48] Samir A, Soliman R, El-Hariri M, Abdel-Moein K, Hatem ME. Leptospirosis in animals and human contacts in Egypt: Broad range surveillance. Revista da Sociedade Brasileira de Medicina Tropical. 2015;**48**(3):272-277

[49] Wagenaar JF, Gasem MH, Goris MG, Leeflang M, Hartskeerl RA, van der Poll T, et al. Soluble ST2 levels are associated with bleeding in patients with severe leptospirosis. PLoS Neglected Tropical Diseases. 2009;**3**(6):e453

[50] Biggs HM, Galloway RL, Bui DM, Morrissey AB, Maro VP, Crump JA. Leptospirosis and human immunodeficiency virus co-infection among febrile inpatients in northern Tanzania. Vector-Borne and Zoonotic Diseases. 2013;**13**(8):572-580

[51] Bruce MG, Sanders EJ, Leake JAD, Zaidel O, Bragg SL, Aye T, et al. Leptospirosis among patients presenting with dengue-like illness in Puerto Rico. Acta Tropica. 2005;**96**(1):36-46

[52] Musso D, La Scola B. Laboratory diagnosis of leptospirosis: A challenge. Journal of Microbiology, Immunology and Infection. 2013;**46**(4):245-252

[53] Abela-Ridder B, Sikkema R, Hartskeerl RA. Estimating the burden of human leptospirosis. International Journal of Antimicrobial Agents. 2010;**36**:S5-S7

[54] Biggs HM, Hertz JT, Munishi OM, Galloway RL, Marks F, Saganda W, et al. Estimating leptospirosis incidence using hospital-based surveillance and a population-based health care utilization survey in Tanzania. PLoS Neglected Tropical Diseases. 2013;**7**(12):e2589

[55] Garba B, Bahaman AR, Bejo SK, Zakaria Z, Mutalib AR, Bande F. Major epidemiological factors associated with leptospirosis in Malaysia. Acta Tropica. 2018;**178**:242-247

[56] Bradley EA, Lockaby G. Leptospirosis and the environment: A review and future directions. Pathogens. 2023;**12**(9):1167

[57] Cristaldi MA, Catry T, Pottier A, Herbreteau V, Roux E, Jacob P, et al. Determining the spatial distribution of environmental and socio-economic suitability for human leptospirosis in the face of limited epidemiological data. Infectious Diseases of Poverty. 2022;**11**(1):86

[58] Bovet P, Yersin C, Merien F, Davis CE, Perolat P. Factors associated with clinical leptospirosis: A population-based case-control study in the Seychelles (Indian Ocean). International Journal of Epidemiology. 1999;**28**(3):583-590

[59] Bierque E, Thibeaux R, Girault D, Soupé-Gilbert ME, Goarant C. A systematic review of Leptospira in water and soil environments. PLoS One. 2020;**15**(1):e0227055

[60] Goarant C. Leptospirosis: Risk factors and management challenges in developing countries. Research and Reports in Tropical Medicine. 2016;**28**:49-62

[61] López-Robles G, Córdova-Robles FN, Sandoval-Petris E, Montalvo-Corral M. Leptospirosis at human-animal-environment interfaces in Latin-America: Drivers, prevention, and control measures. Biotecnia. 2021;**23**(3):89-100

[62] Ame MM. Review on Leptospirosis and its public health significance. Veterinary Medicine – Open Journal. 2021;**6**(2):51-59

[63] Puca E, Pipero P, Harxhi A, Abazaj E, Gega A, Puca E, et al. The role of gender in the prevalence of human leptospirosis in Albania. The Journal of Infection in Developing Countries. 2018;**12**(03):150-155

[64] Zhang H, Zhang C, Zhu Y, Mehmood K, Liu J, McDonough SP, et al. Leptospirosis trends in China, 2007-2018: A retrospective observational study. Transboundary and Emerging Diseases. 2020;**67**(3):1119-1128

[65] Matsushita N, Ng CFS, Kim Y, Suzuki M, Saito N, Ariyoshi K, et al. The non-linear and lagged short-term relationship between rainfall and leptospirosis and the intermediate role of floods in the Philippines. PLoS Neglected Tropical Diseases. 2018;**12**(4):e0006331

[66] Joint, F. A. O., & World Health Organization. Joint WHO/FAO Expert Committee on Zoonoses [Meeting Held in Stockholm from 11 to 16 August 1958]: Second Report. Geneva: World Health Organization; 1959

[67] Mackenzie JS, Jeggo M, Daszak P, Richt JA, editors. One Health: The Human-Animal-Environment Interfaces in Emerging Infectious Diseases, Book Series. Current Topics in Microbiology and Immunology. Vol. 366. Berlin: Springer; 2013

[68] Tilahun Z, Reta D, Simenew K. Global epidemiological overview of leptospirosis. International Journal of Microbiological Research. 2013;**4**(1):9-5

[69] Galan DI. A Retrospective Study of the Urban and Rural Epidemiological Patterns and the Occupational Risk Groups of Leptospirosis in Brazil. The George Washington University, ProQuest Dissertations & Theses. 2021:2841

[70] Fleuret S. A Back and Forth between Tourism and Health: From Medical Tourism to Global Health. Science Society and New Technology Series Tourism and Mobility Systems set. John Wiley & Sons; 2022

[71] Pakoa JG, Soupé-Gilbert ME, Girault D, Takau D, Gaviga J, Gourinat AC, et al. High incidence of leptospirosis in an observational study of hospital outpatients in Vanuatu highlights the need for improved awareness and diagnostic capacities. PLoS Neglected Tropical Diseases. 2018;**12**(6):e0006564

[72] World Health Organization. The World Health Report 2003: Shaping the Future. Geneva: World Health Organization; Oct 2003

[73] Lin XA, Chen Y, Yan J. Recombinant multiepitope protein for diagnosis of leptospirosis. Clinical and Vaccine Immunology. 2008;**15**(11):1711-1714

[74] Verma V, Kala D, Gupta S, Kumar H, Kaushal A, Kuča K, et al. Leptospira interrogans outer membrane protein-based nanohybrid sensor for the diagnosis of leptospirosis. Sensors. 2021;**21**(7):2552

[75] Joshi YP, Kim EH, Cheong HK. The influence of climatic factors on the development of hemorrhagic fever with renal syndrome and leptospirosis during the peak season in Korea: An ecologic study. BMC Infectious Diseases. 2017;**17**:1-11

[76] Deshmukh P, Narang R, Jain J, Jain M, Pote K, Narang P, et al. Leptospirosis in Wardha District, Central India—Analysis of hospital based surveillance data. Clinical Epidemiology and Global Health. 2019;**7**(1):102-106

[77] Desai S, van Treeck U, Lierz M, Espelage W, Zota L, Czerwinski M, et al. Resurgence of field fever in a temperate country: An epidemic of leptospirosis among seasonal strawberry harvesters in Germany in 2007. Clinical Infectious Diseases. 2009;**48**(6):691-697

[78] Garvey P, Connell J, O'flanagan D, McKeown P. Leptospirosis in Ireland: Annual incidence and exposures associated with infection. Epidemiology & Infection. 2014;**142**(4):847-855

[79] Zachary JF. Mechanisms of microbial infections. Pathologic Basis of Veterinary Disease. 2017;**132**:241

[80] Mohammed H, Nozha C, Hakim K, Abdelaziz F, Rekia B. Leptospira: Morphology, classification and pathogenesis. Journal of Bacteriology & Parasitology. 2011;**2**(06):1-4

[81] Fouts DE, Matthias MA, Adhikarla H, Adler B, Amorim-Santos L, Berg DE, et al. What makes a bacterial species pathogenic?: Comparative genomic analysis of the genus Leptospira. PLoS Neglected Tropical Diseases. 2016;**10**(2):e0004403

[82] Adler B, Lo M, Seemann T, Murray GL. Pathogenesis of leptospirosis: The influence of genomics. Veterinary Microbiology. 2011;**153**(1-2):73-81

[83] Cordonin C, Turpin M, Bringart M, Bascands JL, Flores O, Dellagi K, et al. Pathogenic Leptospira and their animal reservoirs: Testing host specificity through experimental infection. Scientific Reports. 2020;**10**(1):7239

[84] Evangelista KV, Coburn J. Leptospira as an emerging pathogen: A review of its biology, pathogenesis and host immune responses. Future Microbiology. 2010;**5**(9):1413-1425

[85] Barragan V, Olivas S, Keim P, Pearson T. Critical knowledge gaps in our understanding of environmental cycling and transmission of Leptospira spp. Applied and Environmental Microbiology. 2017;**83**(19):e01190-e01117

[86] Philip N, Garba B, Neela VK. Long-term preservation of Leptospira spp.: Challenges and prospects. Applied Microbiology and Biotechnology. 2018;**102**:5427-5435

[87] Mutoh Y, Koizumi N, Morino E, Hayakawa K, Kato Y, Ohmagari N. Leptospirosis cases in the Tokyo metropolitan area, Japan. Japanese Journal of Infectious Diseases. 2017;**70**(6):669-671

[88] Cunha M, Costa F, Ribeiro GS, Carvalho MS, Reis RB, Nery N Jr, et al. Rainfall and other meteorological factors as drivers of urban transmission of leptospirosis. PLoS Neglected Tropical Diseases. 2022;**16**(4):e0007507

[89] Lau CL, Smythe LD, Craig SB, Weinstein P. Climate change, flooding, urbanisation and leptospirosis: Fuelling the fire? Transactions of the Royal Society of Tropical Medicine and Hygiene. 2010;**104**(10):631-638

[90] Adler B, de la Peña Moctezuma A. *Leptospira* and leptospirosis. Veterinary Microbiology. 2010;**140**(3-4):287-296

[91] Lane AB, Dore MM. Leptospirosis: A clinical review of evidence based diagnosis, treatment and prevention. World Journal of Clinical Infectious Diseases. 2016;**6**(4):61-66

[92] Al-Orry W, Arahou M, Hassikou R, Charof R, Qasmaoui A, Ouanaim C, et al. Serodiagnosis of human leptospirosis by enzyme linked immunosorbent assay (ELISA). Australasian Medical Journal (Online). 2017;**10**(1):30

[93] Schoonman L, Swai ES. Risk factors associated with the seroprevalence of leptospirosis, amongst at-risk groups in and around Tanga city, Tanzania. Annals of Tropical Medicine & Parasitology. 2009;**103**(8):711-718

[94] Faisal SM, McDonough SP, Chang YF. Leptospira: Invasion, pathogenesis and persistence. In: The Pathogenic Spirochetes: Strategies for Evasion of Host Immunity and Persistence. 2012. pp. 143-172

[95] Monahan AM, Miller IS, Nally JE. Leptospirosis: Risks during recreational activities. Journal of Applied Microbiology. 2009;**107**(3):707-716

[96] Chin VK, Lee TY, Lim WF, Ywy WS, Syafinaz AN, Zamberi S, et al. Leptospirosis in human: Biomarkers in host immune responses. Microbiological Research. 2018;**207**:108-115

[97] Bulach D, Adler B. Leptospiral genomics and pathogenesis. Spirochete Biology: The Post Genomic Era. 2018;**415**:189-214

[98] Faine S. *Leptospira* and leptospirosis. USA: CRC Press Inc; 1994

[99] Ren SX, Fu G, Jiang XG, Zeng R, Miao YG, Xu H, et al. Unique physiological and pathogenic features of Leptospira interrogans revealed by whole-genome sequencing. Nature. 2003;**422**(6934):888-893

[100] Wynwood SJ. Advancing leptospirosis diagnostics in human and animal systems [Doctoral dissertation]. University of the Sunshine Coast. 2015

[101] Mwachui MA, Crump L, Hartskeerl R, Zinsstag J, Hattendorf J. Environmental and behavioural determinants of leptospirosis transmission: A systematic review. PLoS Neglected Tropical Diseases. 2015;**9**(9):e0003843

[102] Bharti AR, Nally JE, Ricaldi JN, Matthias MA, Diaz MM, Lovett MA, et al. Leptospirosis: A zoonotic disease of global importance. The Lancet Infectious Diseases. 2003;**3**(12):757-771

[103] Al-orry W, Arahou M, Hassikou R, Quasmaoui A, Charof R, Mennane Z. Leptospirosis: Transmission, diagnosis and prevention. International Journal of Innovation and Applied Studies. 2016;**15**(3):457

[104] Chacko CS, Jayakumar A, Binu SL, Pant RD, Giri A, Chand S, et al. A short review on leptospirosis: Clinical manifestations, diagnosis and treatment. Clinical Epidemiology and Global Health. 2021;**11**:100741

[105] Fraga TR, Carvalho E, Isaac L, Barbosa AS. Leptospira and leptospirosis. In: Molecular medical microbiology. 3rd edition. Academic Press; 2024. pp. 1849-1871

[106] Togal T, Sener A, Yucel N, Demirbilek S, Akgun FS, Aydogan M, et al. Intensive care of a Weil's disease with multiorgan failure. Journal

of Clinical Medicine Research. 2010;**2**(3):145

[107] Murray GL. The molecular basis of leptospiral pathogenesis. Leptospira and Leptospirosis. 2014;**12**:139-185

[108] Ashaiba A, Arun AB, Prasad KS, Tellis RC. Leptospiral sphingomyelinase Sph2 as a potential biomarker for diagnosis of leptospirosis. Journal of Microbiological Methods. 2022;**203**:106621

[109] Charles JC, Jayarajah U, Subasinghe D. Clinical characteristics and outcomes of patients with leptospirosis complicated with acute pancreatitis: A systematic review. Journal of International Medical Research. 2023;**51**(9):03000605231197461

[110] Seguro AC, Andrade L. Pathophysiology of leptospirosis. Shock. 2013;**39**:17-23

[111] Ferrer MF, Scharrig E, Alberdi L, Cedola M, Pretre G, Drut R, et al. Decay-accelerating factor 1 deficiency exacerbates leptospiral-induced murine chronic nephritis and renal fibrosis. PLoS One. 2014;**9**(7):e102860

[112] Bharu K. Risk and prognostic factors for severe leptospirosis in adult population in intensive care unit hospital raja Perempuan Zainab II [Doctoral dissertation]. Universiti Sains Malaysia; 2017

[113] Martins G, Lilenbaum W. Control of bovine leptospirosis: Aspects for consideration in a tropical environment. Research in Veterinary Science. 2017;**112**:156-160

[114] Loureiro AP, Lilenbaum W. Genital bovine leptospirosis: A new look for an old disease. Theriogenology. 2020;**141**:41-47

[115] Aliberti A, Blanda V, Di Marco Lo Presti V, Macaluso G, Galluzzo P, Bertasio C, et al. Leptospira interrogans Serogroup Pomona in a dairy cattle farm in a multi-host Zootechnical system. Veterinary Sciences. 2022;**9**(2):83

[116] Christopher-Hennings J, Erickson GA, Hesse RA, Nelson EA, Rossow S, Scaria J, et al. Diagnostic tests, test performance, and considerations for interpretation. Diseases of Swine. 2019;**3**:75-97

[117] Picardeau M, Bertherat E, Jancloes M, Skouloudis AN, Durski K, Hartskeerl RA. Rapid tests for diagnosis of leptospirosis: Current tools and emerging technologies. Diagnostic Microbiology and Infectious Disease. 2014;**78**(1):1-8

[118] Sykes JE, Gamage CD, Haake DA, Nally JE. Understanding leptospirosis: Application of state-of-the-art molecular typing tools with a one health lens. American Journal of Veterinary Research. 2022;**83**(10):1-8

[119] Trott DJ, Abraham S, Adler B. Antimicrobial resistance in Leptospira, Brucella, and other rarely investigated veterinary and zoonotic pathogens. Microbiology Spectrum. 2018;**6**(4):10-1128

[120] Heymann DL, Dixon M. Infections at the animal/human interface: Shifting the paradigm from emergency response to prevention at source. In: One Health: The Human-Animal-Environment Interfaces in Emerging Infectious Diseases: Food Safety and Security, and International and National Plans for Implementation of One Health Activities. 2013. pp. 207-215

[121] Davis LI. Winter bird census at Xilitla, San Luis Potosí, Mexico. The Condor. 1952;**54**(6):345-355

[122] Smith S, Kennedy BJ, Dermedgoglou A, Poulgrain SS, Paavola MP, Minto TL, et al. A simple score to predict severe leptospirosis. PLoS Neglected Tropical Diseases. 2019;**13**(2):e0007205

[123] Pereira MM, Schneider MC, Munoz-Zanzi C, Costa F, Benschop J, Hartskeerl R, et al. A road map for leptospirosis research and health policies based on country needs in Latin America. Revista Panamericana de Salud Pública. 2017;**41**:e131

[124] Barazzone GC, Teixeira AF, Azevedo BO, Damiano DK, Oliveira MP, Nascimento AL, et al. Revisiting the development of vaccines against pathogenic Leptospira: Innovative approaches, present challenges, and future perspectives. Frontiers in Immunology. 2022;**12**:760291

[125] Matsuo K, Isogai E, Araki Y. Control of immunologically crossreactive leptospiral infection by administration of lipopolysaccharides from a nonpathogenic strain of *Leptospira biflexa*. Microbiology and Immunology. 2000;**44**(11):887-890

[126] Sonrier C, Branger C, Michel V, Ruvoen-Clouet N, Ganiere JP, Andre-Fontaine G. Evidence of cross-protection within *Leptospira interrogans* in an experimental model. Vaccine. 2000;**19**(1):86-94

[127] Grassmann AA, Souza JD, McBride AJA. A universal vaccine against leptospirosis: Are we going in the right direction? Frontiers in Immunology. 2017;**8**:256

[128] Suepaul SM, Carrington CVF, Campbell M, Borde G, Adesiyun AA. Serovars of Leptospira isolated from dogs and rodents. Epidemiology & Infection. 2010;**138**(7):1059-1070

[129] Lucheis SB, Ferreira RS Jr. Ovine leptospirosis in Brazil. Journal of Venomous Animals and Toxins Including Tropical Diseases. 2011;**17**:394-405

[130] Dhewantara P. Spatial Epidemiological Approaches to Monitor and Measure the Risk of Human Leptospirosis. 2020

[131] Trueba G, Zapata S, Madrid K, Cullen P, Haake D. Cell aggregation: A mechanism of pathogenic Leptospira to survive in fresh water. International Microbiology. 2004;**7**(1):35-40

[132] Rabozzi G, Bonizzi L, Crespi E, Somaruga C, Sokooti M, Tabibi R, et al. Emerging zoonoses: The "one health approach". Safety and Health at Work. 2012;**3**(1):77-83

[133] Miyamoto-Shinohara Y, Sukenobe J, Imaizumi T, Nakahara T. Survival of freeze-dried bacteria. The Journal of General and Applied Microbiology. 2008;**54**(1):9-24

[134] Patra KP, Choudhury B, Matthias MM, Baga S, Bandyopadhya K, Vinetz JM. Comparative analysis of lipopolysaccharides of pathogenic and intermediately pathogenic Leptospira species. BMC Microbiology. 2015;**15**:1-11

[135] Billi D, Potts M. Life and death of dried prokaryotes. Research in Microbiology. 2002;**153**(1):7-12

[136] Zuerner RL, Trueba GA. Characterization of IS 1501 mutants of Leptospira interrogans serovar Pomona. FEMS Microbiology Letters. 2005;**248**(2):199-205

[137] Boggild AK, Liles WC. Travel-acquired illnesses associated with fever. The Travel and Tropical Medicine Manual. 2017;**271**:99

[138] Bhargava A, Ralph R, Chatterjee B, Bottieau E. Assessment and initial management of acute undifferentiated fever in tropical and subtropical regions. BMJ. 2018;**363**:145-152

[139] Ashley E, Bern C, Borok M, Brotherton B, Chappuis F, Chung C, et al. Multisystem diseases and infections. Oxford Medical Handbooks. 5th ed. Oxford University Press; 2021:673-766

[140] Munoz-Zanzi C, Groene E, Morawski BM, Bonner K, Costa F, Bertherat E, et al. A systematic literature review of leptospirosis outbreaks worldwide, 1970-2012. Revista Panamericana de Salud Pública. 2020;**44**:e78

[141] Walika M, De Almeida MM, Delgado RC, González PA. Outbreaks following natural disasters: A review of the literature. Disaster Medicine and Public Health Preparedness. 2023;**17**:e444

[142] Khan F, Mahtab M, Ahmad N, Shukla I, Rizvi M, Azam M, et al. Rapid diagnosis of leptospirosis by IgM ELISA in resource poor settings. International Journal of Health Sciences & Research. 2016;**6**(3):73-79

[143] Blessington T, Schenck AP, Levine JF. A systemic review of the frequency of animal leptospirosis in the southern United States and the implications for human health. Southern Medical Journal. 2020;**113**(5):240-249

[144] De Oliveira D, Figueira CP, Zhan L, Pertile AC, Pedra GG, Gusmão IM, et al. Leptospira in breast tissue and milk of urban Norway rats (*Rattus norvegicus*). Epidemiology & Infection. 2016;**144**(11):2420-2429

[145] Stone NE, Hall CM, Ortiz M, Hutton SM, Santana-Propper E, Celona KR, et al. Diverse lineages of pathogenic Leptospira species are widespread in the environment in Puerto Rico, USA. PLoS Neglected Tropical Diseases. 2022;**16**(5):e0009959

[146] Barcellos C, Sabroza PC. The place behind the case: Leptospirosis risks and associated environmental conditions in a flood-related outbreak in Rio de Janeiro. Cadernos De Saude Publica. 2001;**17**:S59-S67

[147] Liverpool J, Francis S, Liverpool CE, Dean GT, Mendez DD. Leptospirosis: Case reports of an outbreak in Guyana. Annals of Tropical Medicine & Parasitology. 2008;**102**(3):239-245

[148] Supe A, Khetarpal M, Naik S, Keskar P. Leptospirosis following heavy rains in 2017 in Mumbai: Report of large-scale community chemoprophylaxis. The National Medical Journal of India. 2018;**31**(1):19-21

[149] Kakalou E, Tsiamis C. Infectious diseases outbreaks following natural disasters: Risk assessment, prevention, and control. In: Emergency Medicine, Trauma and Disaster Management: From Prehospital to Hospital Care and beyond. Cham: Springer International Publishing; 2021. pp. 525-535

[150] Mugabe VA, Inlamea OF, Ali S, Maholela P, Melchior B, Muianga AF, et al. Surveillance for arboviruses and leptospirosis among non-malarial acute febrile illness outpatients in areas affected by cyclones Idai and Kenneth in Mozambique. Frontiers in Tropical Diseases. 2023;**4**:1091545

[151] Mitchell B, Staines BW, Welch D. Ecology of Red Deer: A Research Review Relevant to their Management in Scotland. Institute of Terrestrial Ecology. Natural Environment Research Council. 1977

[152] Jenkins JC. The Politics of Insurgency. The Farm Worker Movement in the 1960s. New York Chichester, West Sussex: Columbia University Press; 1985

[153] Brown RZ, Gorman GW. The occurrence of leptospiral infections in feral rodents in southwestern Georgia. American Journal of Public Health and the Nations Health. 1960;**50**(5):682-688

[154] Childs CW. Ferrihydrite: A review of structure, properties and occurrence in relation to soils. Zeitschrift für Pflanzenernährung und Bodenkunde. 1992;**155**(5):441-448

[155] Levett PN. Systematics of leptospiraceae. Leptospira and Leptospirosis. 2015;**387**:11-20

[156] Constable PD, Hinchcliff KW, Done SH, Grünberg W. Veterinary Medicine: A Textbook of the Diseases of Cattle, Horses, Sheep, Pigs and Goats. Elsevier Health Sciences; Veterinary Medicine: Ed:11. Vol. 1. 2016

[157] Ward T. Good lives and the rehabilitation of offenders: Promises and problems. Aggression and Violent Behavior. 2002;**7**(5):513-528

[158] Forster PM, Smith CJ, Walsh T, Lamb WF, Lamboll R, Hauser M, et al. Indicators of global climate change 2022: Annual update of large-scale indicators of the state of the climate system and human influence. Earth System Science Data. 2023;**15**(6):2295-2327

[159] Wu X, Lu Y, Zhou S, Chen L, Xu B. Impact of climate change on human infectious diseases: Empirical evidence and human adaptation. Environment International. 2016;**86**:14-23

[160] Semenza JC, Rocklöv J, Ebi KL. Climate change and cascading risks from infectious disease. Infectious Diseases and Therapy. 2022;**11**(4):1371-1390

[161] Crimmins EM, Zhang Y, Saito Y. Trends over 4 decades in disability-free life expectancy in the United States. American Journal of Public Health. 2016;**106**(7):1287-1293

[162] Dhewantara PW, Hu W, Zhang W, Yin WW, Ding F, Al Mamun A, et al. Climate variability, satellite-derived physical environmental data and human leptospirosis: A retrospective ecological study in China. Environmental Research. 2019;**176**:108523

[163] Dechet AM, Parsons M, Rambaran M, Mohamed-Rambaran P, Florendo-Cumbermack A, Persaud S, et al. Leptospirosis outbreak following severe flooding: A rapid assessment and mass prophylaxis campaign; Guyana, January-February 2005. PLoS One. 2012;**7**(7):e39672

[164] Patz JA, Frumkin H. Climate change and human health. In: Environmental Health: From Global to Local. 2016. p. 275

[165] Costello EK, Lauber CL, Hamady M, Fierer N, Gordon JI, Knight R. Bacterial community variation in human body habitats across space and time. Science. 2009;**326**(5960):1694-1697

[166] Cook A, Watson J, Van Buynder P, Robertson A, Weinstein P. 10th anniversary review: Natural disasters and their long-term impacts on the health of communities. Journal of Environmental Monitoring. 2008;**10**(2):167-175

[167] Gubler DJ, Reiter P, Ebi KL, Yap W, Nasci R, Patz JA. Climate variability and change in the United States: Potential impacts on vector-and rodent-borne diseases. Environmental Health Perspectives. 2001;**109**(Suppl. 2):223-233

[168] Andre-Fontaine G, Aviat F, Thorin C. Waterborne leptospirosis: Survival and preservation of the virulence of pathogenic *Leptospira* spp. in fresh water. Current Microbiology. 2015;**71**:136-142

[169] Baker HG, Baker I. Amino-acids in nectar and their evolutionary significance. Nature. 1973;**241**(5391):543-545

[170] Perez J, Brescia F, Becam J, Mauron C, Goarant C. Rodent abundance dynamics and leptospirosis carriage in an area of hyper-endemicity in New Caledonia. PLoS Neglected Tropical Diseases. 2011;**5**(10):e1361

[171] Desvars A, Jégo S, Chiroleu F, Bourhy P, Cardinale E, Michault A. Seasonality of human leptospirosis in Reunion Island (Indian Ocean) and its association with meteorological data. PLoS One. 2011;**6**(5):e20377

[172] Nau LH, Obiegala A, Król N, Mayer-Scholl A, Pfeffer M. Survival time of Leptospira kirschneri serovar Grippotyphosa under different environmental conditions. PLoS One. 2020;**15**(7):e0236007

[173] McMichael AJ, Woodruff RE, Hales S. Climate change and human health: Present and future risks. The Lancet. 2006;**367**(9513):859-869

[174] Joshi BK. Conservation and utilization of agro-biodiversity advanced from 1937 to 2017 in Nepal. In: Devkota F, editor. Krishi Sanchar Smarika. Khumaltar, Nepal: Agricultural Information and Communication Center (AICC), MoAD; 2017. pp. 181-208

[175] Sumi A, Telan EFO, Chagan-Yasutan H, Piolo MB, Hattori T, Kobayashi N. Effect of temperature, relative humidity and rainfall on dengue fever and leptospirosis infections in Manila, the Philippines. Epidemiology & Infection. 2017;**145**(1):78-86

[176] Estrela VV, Aroma J, Sroufer R, Raimond K, Intorne AC, Deshpande A, et al. Remote sensing applications in disease mapping and public health analysis. In: Intelligent Healthcare Systems. 1st edition. CRC Press; 2023. pp. 185-202

[177] Herbreteau V, Demoraes F, Khaungaew W, Hugot JP, Gonzalez JP, Kittayapong P, et al. Use of geographic information system and remote sensing for assessing environment influence on leptospirosis incidence, Phrae province, Thailand. International Journal of Geoinformatics. 2006;**2**(4):43-50

[178] Ledien J, Sorn S, Hem S, Huy R, Buchy P, Tarantola A, et al. Assessing the performance of remotely-sensed flooding indicators and their potential contribution to early warning for leptospirosis in Cambodia. PLoS One. 2017;**12**(7):e0181044

## Chapter 3

# Chronic Leptospirosis in Human Patients

*Beatriz Rivas Sánchez*


## Abstract

Leptospirosis can be a subclinical infection or an infection with a variety of symptoms ranging from a simple flu-like illness to severe forms with or without jaundice, multiple organ failure, and death in less than 5 days. Due to the complexity of the clinical picture and the lack of knowledge about leptospirosis, it frequently goes unnoticed. Due the lack of a 100% sensitive test to detect infection in the early phase, some cases are confused with other pathologies. The chronic leptospirosis is widely accepted in animals, but until recently, chronic leptospirosis was denied in humans. It has been reported that, after the acute phase, patients can continue with symptoms for months or years, which have been called sequelae of the disease. The patient can or not continue to present a high antibody titer and viable *Leptospira* in the urine for many years. In some of them, the isolation of *Leptospira interrogans* has been achieved.


**Keywords:** leptospirosis, chronic human leptospirosis, persistent leptospirosis, dark field, long term leptospirosis

## 1. Introduction

Leptospirosis is a cosmopolitan disease that affects more than 180 animal species, including mammals, birds, amphibians, and reptiles. Many of these animals act as reservoirs of the disease [1–3], so the wide distribution of this infection throughout the world is not unusual. Although it is common in temperate areas, it is more common in subtropical and tropical areas. The temperature and humidity in these areas allow greater survival of the etiological agent in contaminated soil and water [1–2, 4–5].

The global burden of leptospirosis is very significant, affecting several regions with a mortality of up to 26.89%. Therefore, this disease has an important impact on public health around the world [6]. However, the great lack of knowledge about it makes its surveillance, control, and detection difficult, especially in those countries that suffer the consequences of global warming and the serious flooding that occurs [6]. Hartskeerl et al. emphasized that the morbidity and mortality of leptospirosis is significantly underestimated due to under-reporting and lack of epidemiological surveillance [4]. Despite being a re-emerging disease, that has an important human public health and veterinary health with a great impact on the world economy, it continues to be neglected due to the difficulty in diagnosing it [4, 5, 7]. Furthermore, the majority of cases reported in leptospirosis database correspond to the best-known severe forms, icteric and hemorrhagic, confirmed by laboratory [4].



IntechOpen

Leptospirosis may be asymptomatic or have mild symptoms as a simple flu-like illness. Frequently, it can affect every organ progressing into a multisystem illness with or without jaundice such as Weil syndrome and pulmonary hemorrhagic fever and cause death in less than 5 days. It presents a broad spectrum of clinical manifestations. It causes a wide range of signs and symptoms, many of which may be mistaken for other diseases such as Lyme disease and other borreliosis, influenza, dengue, hantavirus, hepatitis, yellow fever, malaria, rickettsioses, salmonellosis and other enteric fevers, pyelonephritis, primary, HIV, meningitis [8, 9], cancer [10, 11], psychotic disorders [12–14], among others, including COVID-19 as an associated infection [15].

## 2. Pathogenesis

To understand the pathogenesis of the disease, we must understand how the bacterium interacts with its host in establishing infection and causing leptospirosis. In general, *Leptospira* penetrates the host through mucous membranes, especially the conjunctivae and the oral cavity, and through broken skin and even through softened skin when it has remained under water for a long time. It is a very thin bacterium, spiral-shaped bodies with two endoflagella that wrap around the protoplasmic cylinder, causing a twisting motion. This is similar to a corkscrew, which gives it a great invasive capacity [16, 17]. It has a wide variety of molecules that act at different levels [18], which allows it to cross the extracellular matrix, quickly invade the bloodstream, and distribute to all body tissues. Furthermore, it evades the immune system [19], as well as activates proinflammatory cytokines that play an important role in the inflammatory damage of the disease [20–22]. Some of these virulence factors are binding proteins (Lsa66, Len, LigB, Omp37, Omp47, LigB, Lsa33, Ompl1, LenA, Lip32, etc.) that allow it to adhere to vessel walls, the activation and binding of leukocytes, increasing inflammation, crossing the different tissue layers, activating or inhibiting the host's enzymes, among other activities [18, 20, 23–25]. It has chemotactic proteins (MCPs) that respond to hemoglobin [26–28], transport enzymes (phospholipases, sphingomyelinases) that degrade cell membranes by beta oxidation, hemolysins (tlyABC) that allow it to obtain iron from the heme group, proteases (collagenases, metalloproteases, hyaluronidases, thermolysins) that degrade the extracellular matrix, lipopolysaccharides (LPS) with endotoxic activity producing an inflammatory reaction, injury to the endothelial lining of the blood vessels, microthrombosis, and decrease in blood platelets, which are necessary for coagulation [22, 25, 29–31]. LPS play an essential role in pathogenesis, colonization, and dissemination of *Leptospira*; the evasion of the host's immune system; as well as the degree of damage on the host [6, 32–33]. Although *Leptospira* adopt a fairly linear transmission framework of adhesion and entry into host cells, the exact molecular pathophysiology remains a perplexing enigma to many; even complex metabolic pathways such as cobalamin biosynthesis and free radical detoxification have been reported [6].

In other ways, different populations of pathogenic bacteria, from normal and environmental microbioma, form biofilms by producing a complex matrix that allows them to adhere to different surfaces and provides them with protection [33–35]. Similary, *Leptospira interrogans* is a biofilm-forming pathogen, which allows it to survive in the environment for months [36, 37]. Other studies have reported the presence of aggregated *Leptospira* similar to biofilms within the lumen of the proximal renal tubules in rodents [38–40] and in the vitreous humor in horses [41], which could explain the evasion of the immune system and therefore its persistence in the host and its antibiotic resistance [37, 41–42]. The differences in sensitivity to *Leptospira* between different species of animals do not only depend on genetics, as observed in the differential activation of LPS of human and murine TLR2/TLR4 [21, 40, 43], but also depend on the differences in genetic predisposition between individuals mediated by histocompatibility antigens (MHC) [40]. Another very important aspect in the pathogenesis is the immune system activation via generation of an inflammatory response that can determine the evolution of the disease. The crosstalk between pro-inflammatory and anti-inflammatory interleukins is a complex and dynamic process that influences to eliminate the bacteria. Dysregulation of balance between both interleukins results in paralysis of the immune system, causing sepsis, multiple organ failure, and death [44, 45].

The above indicates that *Leptospira* is a very complex bacterium with a high pathogenic capacity. It has much virulence factors that allow it to invade any organ of the host, evade host defenses, and cause disease. Often, leptospirosis symptoms may be mistaken for other conditions. Leptospirosis is a perplexing conundrum for many, but much is still unknown about the pathophysiology. However, its ability to evade the immune system and adapt to its host by forming biofilms, which protect it from the adverse environment and the action of antibiotics, allows it not only to persist in its host but also to promote chronic infection.

## 3. Chronic leptospirosis

It is now increasingly recognized that many patients are left with chronic post-leptospirosis [21, 45] symptoms that include recurrent migraines, eye pain, uveitis and iridiocyclitis, chronic fatigue, depression, mood swings, obsessive-compulsive disorder, and other disorders including Tourette syndrome, paresis, paralysis, and parainfectious encephalomyelitis. In many cases, patients can improve with repeated courses of antibiotics [12, 46, 47].

The suspicion of the existence of chronic human leptospirosis dates to the last century when Murgetroyd in 1937 described a case of chronic meningitis in a patient with Weil's disease [48]. Later, in 1984, Nicolescu et al. studied four cases, concluding that in certain circumstances human leptospirosis could present as a chronic disease [49].

In the literature, there are multiple studies and case reports that, even when the authors do not mention it directly, show the existence of chronic leptospirosis. Below, we will present some of these works.

In 1986, Kingscote reported two cases of veterinarians who fell ill with leptospirosis presenting with fever, muscle and joint pain, prostration, and severe headache. After taking the clinical history and its correlation with the compatible clinical picture of leptospirosis, they were given treatment with penicillin and doxycycline, responding favorably. The first was infected while handling a *Leptospira* culture. He began with skin and muscle hyperesthesia, weakness, and delirium. Over the next 8 months, he relapsed three times. The severe headache and fatigue continued for two and a half more years. In a relapse, 2 years later, by plate Microagglutination (MAT) low antibody titers of 1:50 were detected. He was treated again with tetracycline with improvement. Case 2 was a veterinarian who began with fever, lassitude, anuria, profuse diarrhea, chills, followed by inability to focus the eyes, temporary loss of consciousness, and lucid but irrational speech. He also presented jaundice, hyperesthesia in the skin and muscles, ecchymosis on the foot, and significant weight loss. He responded well to crystalline penicillin G. He was diagnosed by darkfield observation of *Leptospira* in blood and urine. The MAT and the indirect hemagglutination test (IHA) were negative. He was left with dizziness and weight loss for more than a year [50].

Orrego et al. conducted a study in Colombia on pig farm workers. MAT and growth inhibition (IC) tests were performed at 1:50 and 1:25, respectively, besides direct observation of urine by dark field and its culture. The results were positive in 3.9% to MAT, 9.8% to IC, 31.4% to direct observation, and 60.8% to culture. This indicates that despite being infected with *Leptospira*, and not reporting acute illness, the workers are chronically infected with very low or negative antibodies [51]. The excreting *Leptospira* through the urine, as happens in animals, indicates a carrier state with a high degree of chronic infection [7].

In other works, the detection of antigens by PCR [52] and monoclonal antibodies [53] in cerebrospinal fluid, urine, and aqueous humor have been reported, months or years after the patient recovered from the acute phase.

Avdeeva, in 2003, studied 170 patients who had suffered from leptospirosis due to *Leptospira interrogans* serovar icterohaemorragiae for up to 9 years. During early or late convalescence, they developed damage to the urinary system (67.6%), hepatobiliary (65.3%), cardiovascular (31.8%), nervous system (13.5%), and eyes (7.1%). For 1 year, 86% of the patients had chronic hepatitis, latent phase of chronic renal failure, and myocardiodystrophy. In both the early and late phases, there were nervous and ocular complications. Sixty-six percent had combined complications, with kidney and liver damage being the most frequent at 66%. A very important fact was the activation of endogenous infection foci within 6 months after acute infection. In 19.4%, the process lasted 1.5 years, and 5.3% continued for several years [54].

In Mexico, Velasco et al., 2009, reported the follow-up of seven cases of human leptospirosis diagnosed with acute leptospirosis by MAT and dark field in blood and urine. Six of them were in contact with dogs sick with leptospirosis, and one patient was in contact with rats and opossums that invaded her house. All of them began with an acute fever, myalgia, arthralgia, abdominal pain and distension, asthenia, and adynamia. Some experienced great prostration and petechiae (43%), diarrhea, nausea, vomiting, polyuria, dysenteric and bronchopneumonic symptoms, diaphoresis, and blurred vision (29%). One of them presented severe nonspecific bilateral uveitis, amnestic aphasia, depression, and mental confusion. All were treated with doxycycline for a maximum of 4 months with good success. The patients relapsed months later and again presented some initial signs and symptoms, so treatment was successfully restored for at least 4 more months [55]. The same authors have observed multiple patients with long-term clinical symptoms characterized by headache; low-grade fever; dark fever syndrome; chronic fatigue; hypersomnia; arthralgia; myalgia, especially in the calves, eye pain; lymphadenopathy; anxiety; depression; hematological disorders; ophthalmological and intestinal disorders; hepatorenal syndrome; central nervous system problems including psychological disorders such as obsessive-compulsive disorder, schizophrenia, jealousy, and Tourette syndrome [47]; uveitis and retinopathy [56, 57]; cardiomegaly [58]; and even Leukemia [11]. In these patients, *Leptospira* have been detected in blood, urine, cerebrospinal fluid (CSF), and other fluids by dark field observation, silver impregnation, specific immunostaining by immunohistochemistry [59, 60] and in some cases by electron microscopy in peripheral blood [60], and *L. interrogans* serovar pomona was isolated in 14 individuals [61]. The majority of them presented antibody titers below 1:80 [59–61], as has been reported in chronic disease in animals [7, 62].

Ganoza et al. studied a hyperendemic area of Peru, finding that 38% of the individuals studied had a high rate of seropositivity, and of these, 60% had evidence of recent subclinical infection by presenting antibody titers by MAT 1:800. But what caught attention was that 5% of the individuals, healthy and without evidence of recent infection, were positive for the detection of *Leptospira* in urine using 16S DNA hybridization. The presence of *Leptospira* in the renal tubules of humans demonstrates that inapparent infection is common and leads to the elimination of the organisms for more than a year, so the possibility that a man is a chronic carrier of leptospirosis is not ruled out and contributes to the transmission of the disease [63].

In the Netherlands, Goris et al. studied 225 patients with persistent complications, diagnosed and confirmed as leptospirosis. The study included a period from 1985 to 1993. Sixty-eight of them (30.2%) presented persistent complications. Of these, 80.7% had complications for more than 2 months after the acute infection, and 21.1% remained for more than 24 months. Complications consisted of depression-like symptoms such as extreme fatigue, malaise, and headache, in addition to myalgia and arthralgia that affected mobility, tinnitus, and psychotic complications. The serovars involved were *L. interrogans* serovar icterohaemorrhagiae, *L. iterrogans* serovar grippotyphosa, *L. interrogans* serovar sejroe, and *L. interrogans* serovar pomona [64].

Kurilungin et al. (2017) studied 37 asymptomatic people in an endemic area in Thailand, of which *L. interrogans* was identified by PCR in one of them [65].

### 3.1 Chronic leptospirosis in the nervous system (CNS)

In 1937, Murgetroyd described the first known case of chronic leptospirosis in England. It was a man who worked on a farm where he apparently became infected because there were many rats near the farm. It was initially diagnosed as Weil's disease, when *L. interrogans* serovar icterohaemorrhagiae was found in urine. In the following months, he presented recurrent symptoms of fever without jaundice until it evolved into chronic meningitis. The diagnosis was confirmed by demonstrating the bacteria in the cerebro-spinal fluid by guinea-pig inoculation. The serology was negative [48].

In 2011, Moschini presented the case of a man who became ill after moving cattle to a high area during a flood in Argentina. The patient presented idiopathic acute febrile syndrome, jaundice, and intense and persistent holocranial headache with photophobia and phonophobia, spontaneously limiting itself in 4 weeks. Later he presented sudden dysarthria and right hemiparesis, which was treated as a cerebrovascular accident (CVA). Weeks later he presented a new clinical event characterized by drowsiness and worsening of dysarthria and right hemiparesis. Computed tomography (CT) showed multiple hypodense lesions suggestive of recurrent ischemic stroke receiving specific treatment. Five months later he showed mild dysarthria, right facio-branchio-crural hemiparesis of mild to moderate intensity, spasticity and osteotendinous hyperreflexia in the right upper limb, and right Hoffmann and Babinski signs. MRI showed multiple chronic ischemic lesions with generalized narrowing of the medium- and large-caliber arteries of the anterior and posterior cerebral circulation consistent with vasculitis. After various analyses to rule out infectious agents and systemic immunological diseases, leptospirosis was diagnosed using MAT in serum and was classified as having a triphasic clinical course: initial phase with acute syndrome, the second phase with jaundice and meningitis, and the third phase with isolated and severe involvement of the CNS in the form of meningitis, vasculitis of large arteries, and multiple cerebral infarcts. He responded favorably with specific treatment with intravenous (IV) ceftriaxone for 2 weeks [66].

Wilson et al. studied a 14-year-old boy with severe combined immunodeficiency. He apparently became infected on a trip to Puerto Rico where he swam in a river and in the sea, presenting fever and hematuria suggestive of nephritis associated with *Leptospira*. Over 10 months, various symptoms such as headache, thrombocytopenia, uveitis, conjunctivitis, photophobia, fatigue, abdominal pain, weight loss, weakness, myalgia, nausea, vomiting, psychiatric symptoms, and hydrocephalus were added. The MRI showed abnormalities in the brain as well as the CSF. He underwent various studies including infectious diseases. *Leptospira* were detected in CSF by a new diagnostic test using unbiased next-generation sequencing. Conventional tests for leptospirosis were negative. Specific treatment was administered with full recovery. Subsequently, the serological tests and the PCR were positive. The authors concluded that, in retrospect, this was a case of chronic neuroleptospirosis [67].

### 3.2 Chronic leptospirosis in eyes

The ocular manifestations of leptospirosis appear after the immune phase, that is, when the systemic symptoms disappear. For this reason, health personnel do not relate these symptoms as part of the disease. Sphilberg et al. studied 11 cases with damage to the anterior chamber of the eye, chronic headache, and a history of acute leptospirosis, on average, 22 years ago. After reevaluating the patients, the authors concluded that the symptoms had remained sequelae of the disease [68]. Tenorio et al. subsequently studied 26 patients clinically diagnosed as uveitis due to Toxoplasmosis, of which 10 had positive IgG antibodies for Toxoplasma, indicating a latent infection, and only one was positive for IgM, indicating active toxoplasmosis. Seventy-three percent of patients were positive to the MAT, mostly at low titers. All patients progressed satisfactorily to the specific treatment. The authors concluded that it is important to consider leptospirosis as a differential diagnosis of ocular Toxoplasmosis [69]. In a subsequent study carried out in patients with chronic uveitis, 31.4% positivity for leptospirosis was reported. The authors suggest including the search for *Leptospira* within the test panel for patients with chronic uveitis [56].

In a study carried out by Rathinam et al. in patients with *Leptospira*-positive uveitis, they found that 14.9% of individuals had cataracts from their first consultation, and in 24.1%, the cataracts evolved between 1 and 6 months. In 19.5%, cataracts resolved by spontaneous absorption between 6 and 18 months. The authors concluded that cataracts can be spontaneously absorbed in a significant number of patients and appear to be exclusive to leptospiral uveitis [70]. Later, Rathinam et al. reported six young patients, with a history of having suffered fever for which they received symptomatic treatment. All six attended consultation with a history of poor vision of 15 to 30 days. Upon examination, mature cataracts were observed in only one eye. Four of them underwent surgery, and a histological study was performed on the lens, observing globular degeneration of the fibers, while the controls from foursenile cataracts showed a uniform arrangement of the lens fibers. With this, the authors conclude that rapid cataract maturation is a characteristic of leptospirosis [71].

Arrieta et al. conducted a review on uveitis caused by leptospirosis in 54 international articles selected from more than 10,000 articles found in various databases. Uveitis due to leptospirosis is a pathology that is becoming more relevant day by day. The pathogenesis of uveitis due to leptospirosis is not well known, but it is believed that, due to the low response against *Leptospira* in the eyes, it proliferates rapidly, producing enzymes and toxins that damage the tissue. In some cases, LPS has been found in aqueous humor, inducing congestion and increasing vascular permeability [31]. It has also been associated with rapid progression of cataracts, probably due to an immunological cross-reaction in the response against *Leptospira* lipoproteins with different ocular components of the retina, lenses, and ciliary bodies [71].

### 3.3 Chronic leptospirosis in heart

In Mexico, a 26-year-old patient was studied who was admitted to the hospital with a 3-month history of jaundice, congestive heart failure, grade IV cardiomegaly, and liver damage. Chagas disease was initially suspected and ruled out by laboratory and blood culture. Leptospirosis was also suspected as a co-associated disease, which was confirmed by dark field observation, silver staining, and immunohistochemistry in blood and urine and by MAT at low titers (1:80). By blood culture, the isolation of *L. interrogans* serovar pomona was achieved. The patient died 2 days after diagnosis. At autopsy, the presence of *Trypanosoma cruzi* in the heart was ruled out, and the presence of *Leptospira* was detected in the heart, lung, kidney, and liver by silver staining, immunohistochemistry, and indirect immunofluorescence [58].

This case is interesting because *Leptospira* is described as the causative agent of dilated heart disease and is probably involved in idiopathic heart diseases in which a precise diagnosis is not found. In fatal cases, there is histopathological evidence of damage to the myocardium; although it is not yet clear whether this damage could cause death, it has been reported that myocardial failure may contribute to the death of the patient [72, 73]. On the other hand, the patient already had a chronic disease, which suggests that *Leptospira* could remain in the heart tissue, just as it occurs in the kidneys and eyes [38, 40–41, 43].

### 3.4 Chronic leptospirosis simulating leukemia

Velasco et al. reported two cases that presented bleeding in the skin, mucous, and viscera. The first, an 83-year-old adult with a history of leptospirosis, presented with a long-term, difficult-to-control allergic dermatitis. One year later, he began with pharyngolaryngitis with hemoptoic expectoration, gingivorrhagia, and purpuric lesions in the lower extremities. The patient was diagnosed with chronic leptospirosis. He was hospitalized and treated with high-dose penicillin with a favorable evolution; the dermatitis completely disappeared. Due to continued low hemoglobin and platelets, he was diagnosed with M3 acute myeloblastic leukemia. He was taken off penicillin and treated with chemotherapy, dying 12 hours later. The second case was a 66-year-old adult, who was admitted to the hospital with a diagnosis of anemic and febrile syndrome of 1 month's duration. He was diagnosed with L2 acute lymphoblastic leukemia and was started on specific therapy, with no response. He continued to have fever, oliguria, respiratory failure, and gastrointestinal bleeding. He was reevaluated and diagnosed with chronic leptospirosis; he died a day later. In the autopsy, the diagnosis of leukemia was ruled out in both cases. *Leptospira* were found in bone marrow, kidneys, spleen, liver, lungs, and digestive tract, stained by Warthin Starry and indirect immunofluorescence with specific antibodies. The histopathological diagnoses were generalized leptospirosis with spinal cord aplasia and generalized leptospirosis with myelodysplastic syndrome, respectively [11].

### 3.5 Chronic leptospirosis in kidney

Chronic kidney disease (CKD) is a state of progressive loss of kidney function caused by damage from a pathological process such as hypertension, diabetes, infection, and so on. The incidence of CKD is difficult to determine because it often goes unnoticed until CKD is in an advanced stage [74]. It is estimated that it affects 10% of the world's population, that is, more than 800 million people. CKD is a leading non-communicable cause of death that has been increasing over the past two decades [75].

The kidney is a very important target organ in leptospirosis, and kidney damage is an early manifestation of acute leptospirosis called acute renal failure (ARF). The ARF involves acute tubular necrosis and interstitial nephritis with tubular degeneration and interstitial edema that can be severe and diffuse. The incidence of ARF varies from 10 to 60%, depending on the age of the patient, the severity of the disease, and the infecting serovar [76]. Recovery of kidney function can take months or years, during which the individual may remain an asymptomatic carrier [63, 77, 78]. It has also been observed that many patients who develop severe ARF may continue to present antibody titers and viable *Leptospira* in urine for years, especially in endemic areas, which strongly supports that *Leptospira* causes chronic kidney disease [78–83].

In 2014 in Sri Lanka, a one-year follow-up study of 44 patients who suffered from ARF due to acute leptospirosis was reported. Of these 31, they recovered their kidney function in a few days. Eleven presented a decrease in their glomerular filtration rate and an increase in creatinine, the measurement of which represents a marker of kidney function. During the follow-up period, seven patients recovered their kidney function, and four showed damage compatible with CKD. The authors conclude that 9% of patients showed altered renal function consistent with an early stage of CKD [80].

Huang et al. conducted a study of 3045 chronic kidney patients at different stages of evolution in an area with a high dialysis rate in Taiwan. It was observed that subjects with prior *Leptospira* exposure had a lower level of glomerular filtration rate (eGFR) and more severe CKD disease, especially in subjects with high antibody titers with an elevated level of KIM-1 compared to those individuals with low antibodies. The KIM-1 is a biomarker associated with kidney disease. Two participants with persistently high antibody titers and impaired renal function were positive for PCR in urine, demonstrating the presence and elimination of *Leptospira* through urine. These results suggest that prior exposure to *Leptospira* is associated with a higher prevalence and severity of CKD [81].

Chih relate the endemic areas of leptospirosis with the areas with the highest number of cases of CKD of unknown origin. Many of these patients report having suffered symptoms like the acute phase of leptospirosis without it having been diagnosed in a timely manner, due to the lack of knowledge and adequate diagnostic tests. Both in these patients and the asymptomatic, it is likely that the bacteria have not been completely eradicated and that the individuals have progressed to CKD. Chronic leptospirosis may develop if *Leptospira* persists in the tubular lumen and interstitium as a continuum of acute kidney injury or asymptomatic renal colonization with clinical symptoms occurring in 2.3–57% of the population at risk. Chronic leptospiral renal infection may present with characteristic chronic tubulointerstitial nephritis and interstitial fibrosis, a well-known feature in animals. *Leptospira* can colonize the proximal tubules and progress to chronic tubulointerstitial nephritis and fibrosis, so asymptomatic exposure may be associated with an increased risk of developing CKD [82].

Later, Chih et al. conducted a retrospective study in 2145 patients diagnosed with acute leptospirosis. Of these, 443 (20.6%) developed ARF. No significant differences in mortality risk were observed between the ARF group and the non-ARF group. In total 282 (13.14%) patients developed CKD, being significantly higher in patients with ARF, especially patients who were dialyzed. The authors conclude that leptospirosis patients complicated with ARF are associated with a higher risk of CKD [83].

## 4. Diagnosis and treatment

Leptospirosis is a zoonotic disease of global public health importance. However, due to the lack of knowledge about it and the confusion on protean clinical picture, it is often not taken into consideration by health professionals. It is difficult to make a precise diagnosis, even in the acute phase, despite the serious or fatal consequences. Just as occurred in the Nicaragua outbreak in 1995, where the serology tests were unable to detect leptospirosis, the diagnosis was made in postmortem histopathological studies that were conducted on the first deceased, when *Leptospira* were observed in the different tissue samples stained with silver impregnation and immunohistochemistry [84]. Similarly, in India, 62 autopsies of individuals clinically suspected of leptospirosis were analyzed. The diagnosis was confirmed postmortem in 39 cases (63%) by immunohistochemistry and Levaditi's staining of the kidney and serology by MAT [85].

In chronic leptospirosis, the diagnosis is more difficult to make. Frequently, it is a diagnosis by rule out when all clinical suspicions of better-known diseases are negative. Velasco et al. propose that the diagnosis of both acute and chronic leptospirosis should be made sequentially or *"in construct"* [59]. Diagnostic constructs are created by the interaction of causal factors, multiple cultural and individual factors that give rise to a medical condition. Therefore, diagnostic constructs based on similarity with the clinical picture can have clinical utility by accumulating information about probable results [86], where the collection, integration, and interpretation of information, through the preparation of the clinical history, is essential to reach the diagnosis.

Laboratory diagnosis of chronic leptospirosis should be performed by searching for antibodies at low titers, from a dilution of 1:20 [50, 51, 59, 61, 62]. Likewise, in addition to serology for diagnosis, it is necessary to use two or more test to detect the bacteria in different tissues, including blood and urine. The use of dark field observation is very controversial, but if it is performed by specialized personnel, it is an easy and quick diagnostic test to perform. **Figure 1** shows stain with silver salts such as Whartin Starry and Fontana, which are useful to detect the bacteria in different tissues; the sensitivity of this test is increased when the observation is made under a darkfield microscope. **Figure 2** shows the use of specific antibodies against *Leptospira* by immunostaining, which is a very useful and sensitive test for diagnosis, such as indirect immunofluorescence and immunohistochemistry, as shown in **Figure 3** [47, 59]. In recent years, serological and immunohistochemical tests have been developed using specific recombinant antigens with good expectations for diagnosis [87]. Molecular tests such as PCR are very useful tests for diagnosis since they are very specific tests, but their sensitivity depends on the type of sample, the gene, and the molecular method used [87–89].

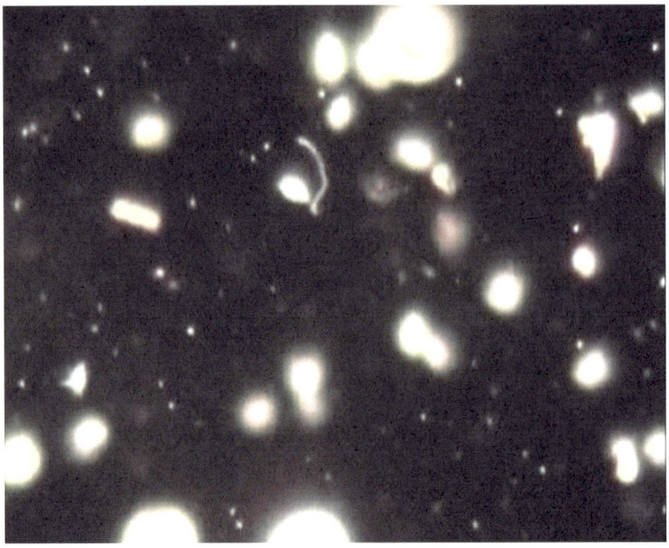

**Figure 1.**
Leptospira *in blood observed by dark field microscopy.*

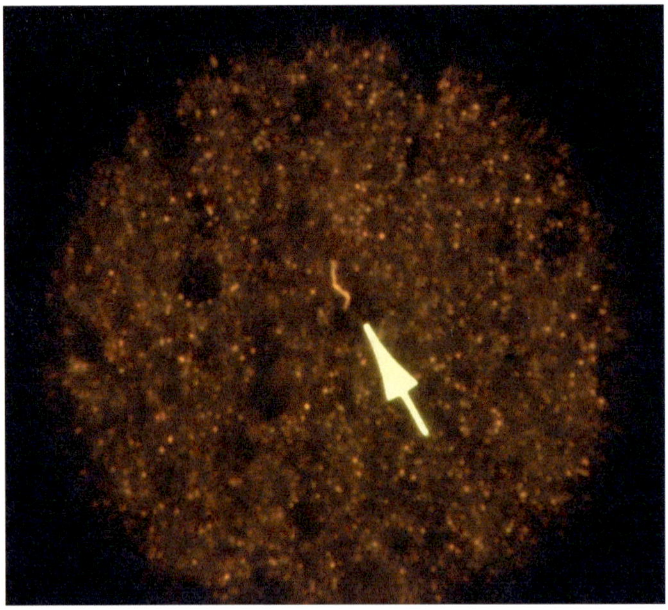

**Figure 2.**
Leptospira *with silver stain (Whartin Starry) observed by dark field microscopy.*

Regarding treatment, the same antibiotics used in the acute phase are used in the treatment of chronic leptospirosis [90], the base antibiotics being penicillin and doxycycline [91]. Since it is a bacterium that can survive in the organs for many years, courses of antibiotics must be given for a long time [12, 46, 47], even for more than a year, and the patient's immune system must be strengthened with immunomodulators such as transfer factor [61, 92].

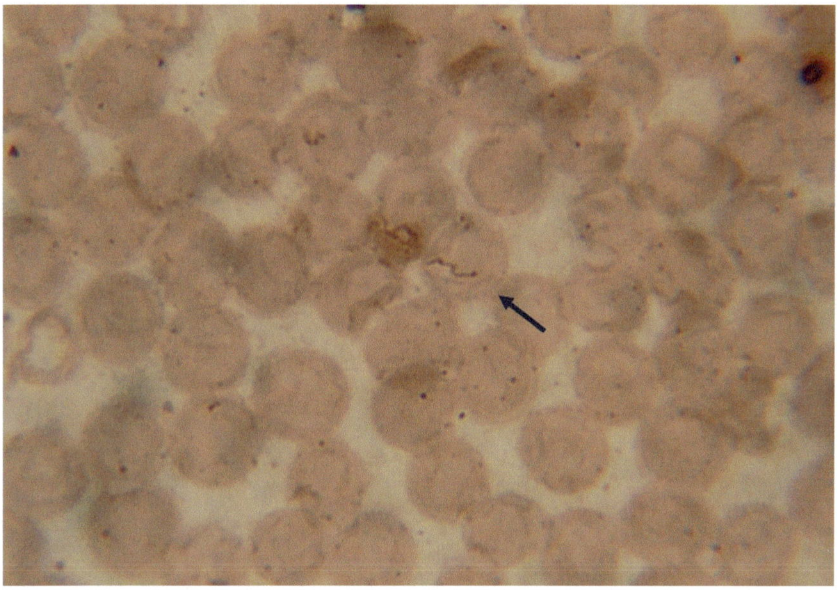

**Figure 3.**
*Immunohistochemestry showing* Leptospira interrogans.

## 5. Conclusion

Chronic leptospirosis disease occurs with a very complex clinical picture that can be misdiagnosed with other diseases or go unnoticed due to the lack of recognition of the symptoms [4–7]. When patients are diagnosed and treated for acute leptospirosis, many of them continue to have symptoms months or years after the infection has subsided [55, 64]. Although it is said that these are sequelae left by disease [68], there are multiple reports that indicate that it is a long-term, chronic disease, and the patient's health status improves after antibiotic treatment [12, 46–48].

Many of these chronic cases present antibody titers starting at 1:20 [56, 65–67]. The presence of *Leptospira* is observed in blood, urine, and other biological samples by dark field microscopy [50–51, 58–61], by silver staining, and by immunohistochemistry with specific antibodies [58–60, 84–85]. The isolation of *Leptospira* from blood, urine, cerebro-spinal fluid, and biopsies of heart and skin is also reported [48, 50, 51, 58, 61].

These findings confirm the existence of chronic leptospirosis, which, until recently, had been denied worldwide. In some epidemiological studies, *Leptospira* has even been detected in blood and the urine of asymptomatic individuals living in endemic areas [93, 94], in hyperendemic areas [63], or in high-risk workers [51], confirming that *Leptospira* can survive in humans for a long time, just as it does in animals, without causing any symptoms.

Acute leptospirosis is one of the most important zoonotic diseases in the world, which makes it a serous global public health problem. But the growing recognition of the existence of chronic human leptospirosis contributes to a significant increase in its global burden [64]. All of these findings support the confirmation of the existence of chronic leptospirosis, which until recently was denied worldwide.

## Author details

Beatriz Rivas Sánchez
An Independent Researcher, Mexico City, Mexico

*Address all correspondence to: beatriz_rivass@yahoo.com.mx

IntechOpen

© 2025 The Author(s). Licensee IntechOpen. This chapter is distributed under the terms of the Creative Commons Attribution License (http://creativecommons.org/licenses/by/3.0), which permits unrestricted use, distribution, and reproduction in any medium, provided the original work is properly cited.

## References

[1] Levett PN. Leptospirosis. Clinical Microbiology Reviews. 2001;**14**(2):296-326. DOI: 10.1128/CMR.14.2.296-326.2001

[2] Bharti AR, Nally JE, Ricaldi JN, Matthias MA, Diaz MM, Lovett MA, et al. Leptospirosis: A zoonotic disease of global importance. The Lancet Infectious Diseases. 2003;**3**(12):757-771. DOI: 10.1016/s1473-3099(03)00830-2. Available from: https://pubmed.ncbi.nlm.nih.gov/14652202/

[3] Vieira AS, Pinto P, Lilenbaum W. A systematic review of leptospirosis on wild animals in Latin America. Tropical Animal Health and Production. 2018;**50**(2):229-238. DOI: 10.1007/s11250-017-1429-y. Available from: https://pubmed.ncbi.nlm.nih.gov/28967042/

[4] Hartskeerl RA, Collares-Pereira M, Ellis WA. Emergence, control and re-emerging leptospirosis: Dynamics of infection in the changing world. Clinical Microbiology and Infection. 2011;**17**(4):494-501. DOI: 10.1111/j.1469-0691.2011.03474.x. Available from: https://pubmed.ncbi.nlm.nih.gov/21414083/

[5] World Health Organization. Report of the First Meeting of the Leptospirosis Burden Epidemiology Reference Group. Geneva: World Health Organization; 2010. ISBN 9789241599894. Available from: https://iris.who.int/handle/10665/44382

[6] Samrot AV, Sean TC, Bhavya KS, Sahithya CS, Chan-Drasekaran S, Palanisamy R, et al. Leptospiral infection, pathogenesis and its diagnosis-A review. Pathogens. 2021;**10**(2):145. DOI: 10.3390/pathogens10020145. Available from: https://pubmed.ncbi.nlm.nih.gov/33535649/

[7] World Organization for Animal Health. Leptospirosis. In: Manual de las pruebas de diagnóstico y de las vacunas para los animales terrestres. París: OIE (WOAH); 2021. Capítulo 3.1.12. Available from: https://www.woah.org/fileadmin/Home/eng/Health_standards/tahm/3.01.12_LEPTO.pdf

[8] WHO Team. Human Leptospirosis: Guidance for Diagnosis, Surveillance and Control. Republic of Malta: Malta WHO; 2003. ISBN 9241545895. Available from: https://www.who.int/publications/i/item/human-leptospirosis-guidance-for-diagnosis-surveillance-and-control

[9] Silesky JJI, Hidalgo Marroquin JL, Richards GA, Amin P. Leptospirosis: Report from the task force on tropical diseases by the world Federation of Societies of intensive and critical care medicine. Journal of Critical Care. 2018;**43**:361-365. DOI: 10.1016/j.jcrc.2017.11.005. Available from: https://pubmed.ncbi.nlm.nih.gov/29129539/

[10] Granito A, Ballardini G, Fusconi M, Volta U, Muratori P, Sambri V, et al. A case of leptospirosis simulating colon cancer with liver metastases. World Journal of Gastroenterology. 2004;**10**(16):2455-2456. DOI: 10.3748/wjg.v10.i16.2455. Available from: https://pubmed.ncbi.nlm.nih.gov/15285043/

[11] Velasco-Castrejón O, Rivas-Sánchez B, Gutierrez E, Chávez L, Duarte P, Chavarría S, et al. Leptospira: ¿simulador o causante de leucemia? Revista Cubana de Medicina Tropical. 2005;**57**:17-24. ISSN 0375-0760. Available from: http://scielo.sld.cu/pdf/mtr/v57n1/mtr03105.pdf

[12] The Leptospirosis Information Center. Persistent Human Leptospirosis. United Kingdom: LIC website; 2009. Available from: https://www.leptospirosis.org/medical/phl.php

[13] Semiz UB, Turhan V, Basoglu C, Oner O, Ebrinc S, Cetin M. Leptospirosis presenting with mania and psychosis: Four consecutive cases seen in a military hospital in Turkey. International Journal of Psychiatry in Medicine. 2005;**35**(3):299-305. DOI: 10.2190/0KUX-JE1J-HV6K-3R9J. Available from: https://pubmed.ncbi.nlm.nih.gov/16480244/

[14] Craig SB, Prior SJ, Weier SL, Graham GC, Collet TA, Moore FAJ, et al. Leptospirosis: Messing with our minds- A review of unusual neurological and psychiatric complexities. In: Sing A, editor. Zoonoses: Infections Affecting Humans and Animals. Cham: Springer; 2023. pp. 1313-1330. DOI: 10.1007/978-3-031-27164-9_34

[15] Nakandakari MD, Marín-Macedo H, Seminario-Vilca R. Dengue with signs of alarm and leptospirosis in a pediatric patient with Covid-19. Revista de la Facultad de Medicina Humana. 2021;**21**(2):438-444. Available from: www.scielo.org.pe/scielo.php?pid=S2308-05312021000200438&script=sci_abstract&tlng=en

[16] Cameron CE. Leptospiral structure, physiology, and metabolism. In: Adler B, editor. Leptospira and Leptospirosis. Current Topics in Microbiology and Immunology. Vol. 387. Berlin, Heidelberg: Springer; 2014. pp. 21-41. DOI: 10.1007/978-3-662-45059-8_3

[17] Picardeau M. Virulence of the zoonotic agent of leptospirosis: Still terra incognita? Nature Rew Microbiol. 2017;**15**(5):297-307 Available from: https://pubmed.ncbi.nlm.nih.gov/28260786/

[18] Haake DA, Matsunaga J. Leptospira: A spirochaete with a hybrid outer membrane. Molecular Microbiology. 2010;**77**(4):805-814. DOI: 10.1111/j.1365-2958.2010.07262.x. Available from: https://www.ncbi.nlm.nih.gov/pmc/articles/PMC2976823/

[19] Mohammed H. Leptospira: Morphology, classification and pathogenesis. Journal of Bacteriology & Parasitology. 2011;**2**(6):2-6. Available from: https://www.researchgate.net/publication/234163768_LEPTOSPIRA_Morphology_Classification_and_Pathogenesis

[20] Cinco M. New insights into the pathogenicity of leptospires: Evasion of host defences. The New Microbiologica. 2010;**33**(4):283-292. Available from: https://pubmed.ncbi.nlm.nih.gov/21213586/

[21] Haake DA, Levett PN. Leptospirosis in humans. Current Topics in Microbiology and Immunology. 2015;**387**:65-97. DOI: 10.1007/978-3-662-45059-8_5. Available from: https://pubmed.ncbi.nlm.nih.gov/25388133/

[22] Ai-Hua S, Xiao-Xiang L, Jie Y. Leptospirosis is an invasive infectious and systemic inflammatory disease. Biomedical Journal. 2020;**43**(1):24-31. DOI: 10.1016/j.bj.2019.12.002. Available from: https://www.ncbi.nlm.nih.gov/pmc/articles/PMC7090314/

[23] Yi-Pin L, McDonough SP, Sharma Y, Yung-Fu C. The terminal immunoglobulin-like repeats of LigA and LigB of *Leptospira* enhance their binding to Gelatin binding domain of fibronectin and host cells. PLoS One. 2010;**5**(6):e11301. DOI: 10.1371/journal.pone.0011301

[24] Stevenson B, Choy HA, Pinne M, Rotondi ML, Miller MC, DeMoll E, et al. *Leptospira interrogans* Endostatin-like outer membrane proteins bind host fibronectin, Laminin and regulators of complement. PLoS One. 2007;**2**(11):e1188. DOI: 10.1371/journal.pone.0001188. Available from: https://pubmed.ncbi.nlm.nih.gov/18000555/

[25] Daroz BB, Fernandes LGV, Cavenague MF, Kochi LT, Passalia FJ, Takahashi MB, et al. A review on host-*Leptospira* interactions: What we know and future expectations. Frontiers in Cellular and Infection Microbiology. 2021;**11**:777709. DOI: 10.3389/fcimb.2021.777709. Available from: https://pubmed.ncbi.nlm.nih.gov/34900757/

[26] Yuri K, Takamoto Y, Okada M, Hiramune T, Kikuchi N, Yanagawa R. Chemotaxis of leptospires to hemoglobin in relation to virulence. Infection and Immunity. 1993;**61**(5):2270-2272. DOI: 10.1128/iai.61.5.2270-2272.1993. Available from: https://www.ncbi.nlm.nih.gov/pmc/articles/PMC280839/pdf/iai00017-0686.pdf

[27] Ko AI, Goarant C, Picardeau M. *Leptospira*: The dawn of the molecular genetics era for an emerging zoonotic pathogen. Nature Reviews. Microbiology. 2009;**7**(10):736-747. DOI: 10.1038/nrmicro2208. Available from: https://www.ncbi.nlm.nih.gov/pmc/articles/PMC3384523/pdf/nihms318803.pdf

[28] Santos JC, Vieira ML, Abendroth J, Lin T, Staker BL, Myler PJ, et al. Structural analysis of CACHE domain of the McpA chemoreceptor from *Leptospira interrogans*. Biochemical and Biophysical Research Communications. 2020;**533**(4):1323-1329. DOI: 10.1016/j.bbrc.2020.10.013. Available from: https://www.ncbi.nlm.nih.gov/pmc/articles/PMC7744396/

[29] Boza R. Sobre la patogénesis de la leptospirosis. Revista Costarricense de Ciencias Médicas. 1999;**20**(1-2):115-120. Available from: http://www.scielo.sa.cr/scielo.php?script=sci_arttext&pid=S0253-29481999000100011&lng=en

[30] Patra KP, Choudhury B, Matthias MM, Baga S, Bandyopadhya K, Vinetz JM. Comparative analysis of lipopolysaccharides of pathogenic and intermediately pathogenic Leptospira species. BMC Microbiology. 2015;**15**:244-254. DOI: 10.1186/s12866-015-0581-7. Available from: https://pubmed.ncbi.nlm.nih.gov/26518696/

[31] Arrieta-Bechara CE, Carrascal-Maldonado AY. Ocular leptospirosis: A review of current state of art of a neglected disease. Romanian Journal of Ophthalmology. 2022;**66**(4):282-288. DOI: 10.22336/rjo.2022.53. Available from: https://pubmed.ncbi.nlm.nih.gov/36589326/

[32] Fraga TR, Barbosa AS, Isaac L. Leptospirosis: Aspects of innate immunity, Immunopathogenesis and immune evasion from the complement system. Scandinavian Journal of Immunology. 2011;**73**:408-419. Available from: https://pubmed.ncbi.nlm.nih.gov/21204903/

[33] Cheng KJ, Costerton JW. Ultrastructure of cell envelopes of bacteria of the bovine rumen. Applied Microbiology. 1975;**29**(6):841-849. DOI: 10.1128/am.29.6.841-849.1975. Available from: https://pubmed.ncbi.nlm.nih.gov/1171655/

[34] Costerton JW, Irvin RT, Cheng KJ. The bacterial Glycocalyx in nature and disease. Annual Review of Microbiology. 1981;**35**:299-324. DOI: 10.1146/annurev.mi.35.100181.001503. Available from: https://pubmed.ncbi.nlm.nih.gov/7027902/

[35] Lappin-Scott H, Burton S, Stoodley P. Revealing a world of biofilms — The pioneering research of Bill Costerton. Nature Reviews. Microbiology. 2014;**12**:781-787. DOI: 10.1038/nrmicro3343

[36] Thibeaux R, Soupé-Gilbert ME, Kainiu M, Girault D, Bierque E, Fernandez J, et al. The zoonotic pathogen *Leptospira interrogans* mitigates environmental stress through cyclic-di-GMP-controlled biofilm production. npj Biofilms and Microbiomes. 2020;**6**:24. DOI: 10.1038/s41522-020-0134-1. Available from: https://www.nature.com/articles/s41522-020-0134-1

[37] Rezende Mde CR, Silva DC, Nogueira Paz L, Melo de LFT, Pereira FC, Araújo DK, et al. Biofilm formation *in vitro* by *Leptospira interrogans* strains isolated from naturally infected dogs and their role in antimicrobial resistance. Heliyon. 2023;**9**(3):e13802. DOI: 10.1016/j.heliyon.2023.e13802. Available from: https://www.ncbi.nlm.nih.gov/pmc/articles/PMC9976318/

[38] Sterling CR, Thiermann AB. Urban rats as chronic carriers of leptospirosis: An ultrastructural investigation. Veterinary Pathology. 1981;**18**(5):628-637. DOI: 10.1177/030098588101800508

[39] Ristow P, Bourhy P, Kerneis S, Schmitt C, Prevost MC, Lilenbaum W, et al. Biofilm formation by saprophytic and pathogenic leptospires. Microbiology. 2008;**154**:1309-1317. DOI: 10.1099/mic.0.2007/014746-0. Available from: https://pubmed.ncbi.nlm.nih.gov/18451039/

[40] Monahan AM, Callanan JJ, Nally JE. Review paper: Host-Pathogen interactions in the kidney during chronic leptospirosis. Veterinary Pathology. 2009;**46**:792-799. DOI: 10.1354/vp-08-VP-0265-N-REV. Available from: https://journals.sagepub.com/doi/epub/10.1354/vp.08-VP-0265-N-REV

[41] Ackermann K, Kenngott R, Settles M, Gerhards H, Maierl J, Wollanke B. In vivo biofilm formation of pathogenic *Leptospira* spp. in the vitreous humor of horses with recurrent uveitis. Microorganisms. 2021;**9**(9):1915. DOI: 10.3390/microorganisms9091915. Available from: https://www.ncbi.nlm.nih.gov/pmc/articles/PMC8464839/

[42] Ratet G, Veyrier FJ, Fanton D'AM, Kammerscheit X, Nicola MA, Picardeau M, et al. Live imaging of bioluminescent *Leptospira interrogans* in mice reveals renal colonization as a stealth escape from the blood Defenses and antibiotics. PLoS Neglected Tropical Diseases. 2014;**8**(12):e3359. DOI: 10.1371/journal.pntd.0003359. Available from: https://journals.plos.org/plosntds/article?id=10.1371/journal.pntd.0003359

[43] Nahori MA, Fournié-Amazouz E, Que-Gewirth NS, Balloy V, Chignard M, Raetz CRH, et al. Differential TLR recognition of leptospiral lipid A and lipopolysaccharide in murine and human cells. Journal of Immunology. 2005;**175**(9):6022-6031. DOI: 10.4049/jimmunol.175.9.6022

[44] Cagliero J, Villanueva SYAM, Matsui M. Leptospirosis pathophysiology: Into the storm of cytokines. Frontiers in Cellular and Infection Microbiology. 2018;**8**:204. DOI: 10.3389/fcimb.2018.00204. Available from: https://pubmed.ncbi.nlm.nih.gov/29974037/

[45] Chancharoenthana W, Leelahavanichkul A, Schultz MJ, Dondorp AM. Going micro in leptospirosis kidney disease. Cells. 2022;**11**(4):698. DOI: 10.3390/cells11040698

[46] Izurieta R, Galwankar S, Clem A. Leptospirosis: The "mysterious" mimic. Journal of Emergencies, Trauma, and Shock. 2008;**1**(1):21-33. DOI: 10.4103/0974-2700.40573. Available from: https://www.ncbi.nlm.nih.gov/pmc/articles/PMC2700559/

[47] Velasco-Castrejon O, Rivas-Sánchez B, Rivera-Reyes HH. Leptospirosis humana crónica. In: Narro-Robles J, Rivero SO, Lopez BJJ, editors. Diagnostico y Tratamiento en la Práctica Médica. El Manual Moderno. 3rd ed. México DF: El Manual Moderno; 2008. pp. 765-775. ISBN: 9707293276

[48] Murgatroyd F. Chronic meningitis in Weil's disease. British Medical Journal. 1937;**1**:7-11 Available from: brmedj04231-0008.pdf (nih.gov)

[49] Nicolescu M, Andreescu N. May human leptospirosis develop as a chronic infection? Zentralblatt für Bakteriologie, Mikrobiologie, und Hygiene. Series A. 1984;**257**(4):531-534. Available from: https://pubmed.ncbi.nlm.nih.gov/6506925/

[50] Kingscote BF. Leptospirosis: An occupational Hazard to veterinarians. The Canadian Veterinary Journal. 1986;**27**:78-81

[51] Orrego UA, Giraldo deLG, Rios AB, Valencia PPA. Leptospirosis en personas de riesgo de quince explotaciones porcinas y de la central de sacrificio de Manizales, Colombia. Archivos de Medicina Veterinaria. 2003;**35**(2):205-213. DOI: 10.4067/S0301-732X2003000200008

[52] Romero EC, Billerbeck AE, Lando VS, Camargo ED, Souza CC, Yasuda PH. Detection of *Leptospira* DNA in patients with aseptic meningitis by PCR. Journal of Clinical Microbiology. 1998;**36**(5):1453-1455. Available from: https://www.ncbi.nlm.nih.gov/pmc/articles/PMC104853/pdf/jm001453.pdf

[53] Saengjaruk P, Chaicumpa W, Watt G, Bunyaraksyotin G, Wuthiekanun V, Tapchaisri P, et al. Diagnosis of human leptospirosis by monoclonal antibody-based antigen detection in urine. Journal of Clinical Microbiology. 2002;**40**(2):480-489. DOI: 10.1128/JCM.40.3.480-489.2002. Available from: https://www.ncbi.nlm.nih.gov/pmc/articles/PMC153370/pdf/0766.pdf

[54] Avdeeva MG. Outcome and tendency of late convalescence in icterohemorrhagic leptospirosis. Clinical Medicine (Russian Journal). 2003;**81**:42-47

[55] Velasco-Castrejón O, Rivas-Sánchez B, Rivera RHH. Transición de la leptospirosis aguda a crónica. Seguimiento de siete casos. Revista Mexicana de Patología Clínica y Medicina de Laboratorio. 2009;**56**:183-192

[56] Elizalde CAA, Tenorio GG, Velasco CO. Identificación de *leptospira* en la patogénesis de la uveítis crónica en la ciudad de México. Revista Mexicana de Oftalmología. 2004;**78**(4):165-170. Available from: https://www.medigraphic.com/pdfs/revmexoft/rmo-2004/rmo044b.pdf

[57] Velasco-Castrejón O, Rivas-Sánchez B, Tenorio GG. Diagnóstico diferencial entre Toxoplasmosis y Leptospirosis de 26 casos mexicanos de uveitis posterior. Revista Cubana de Medicina Tropical. 2005;**57**(1):77

[58] Velasco-Castrejón O, Rivas-Sánchez B, Soriano RJ, Rivera-Reyes HH. Daño miocárdico grave por leptospirosis. Informe de un caso fatal en México. Archivos de Cardiología de México. 2009;**79**(4):268-273. Available from: https://www.elsevier.es/

es-revista-archivos-cardiologia-mexico-293-pdf-X1405994009476432

[59] Velasco-Castrejón O, Rivas-Sánchez B, Espinoza HJ, Martínez HE. Diagnóstico de leptospirosis crónica, comparación entre a aglutinación microscópica y 3 técnicas diagnósticas confirmatorias. Revista Cubana de Medicina Tropical. 2007;**59**(1):8-13. Available from: https://www.researchgate.net/publication/262721909_Diagnostico_de_leptospirosis_cronica_comparacion_entre_la_aglutinacion_microscopica_y_3_tecnicas_diagnosticas_confirmatorias

[60] Velasco-Castrejón O, Rivas-Sánchez B, Sánchez SME, Soriano J, Rivera- Reyes HH, Garibay-Sablez V. Leptospirosis crónica en México: diagnóstico microscópico y evidencias que respaldan su existencia e importancia. Revista Mexicana de Patología Clínica y Medicina de Laboratorio. 2009;**56**(3):157-167. Available from: https://www.medigraphic.com/pdfs/patol/pt-2009/pt093a.pdf

[61] Rivas-Sánchez B, Velasco-Castrejón O, Jimenez MJ. Isolation of *L. interrogans* Serovar Pomona in 14 human cases and an African lion, all with chronic leptospirosis. Open. Journal of Medical Microbiology. 2016;**6**:158-170. Available from: https://www.scirp.org/pdf/OJMM_2016122115470954.pdf

[62] Centro Panamericano de Zoonosis. Procedimientos para estudio de prevalencia de enfermedades crónicas en el ganado. OPS/OMS. Nota técnica No. 18, Buenos Aires Argentina. 1973. 35 p

[63] Ganoza CA, Matthias MA, Saito M, Cespedes M, Gotuzzo E, Vinetz JM. Asymtomatic renal colonization of humans in the Peruvian Amazon by *Leptospira*. PLoS Neglected Tropical Diseases. 2010;**4**:e612. DOI: 10.1371/journal.pntd.0000612

[64] Goris MG, Kikken V, Straetemans M, Alba S, Goeijenbier M, van Gorp EC, et al. Towards the burden of human leptospirosis: Duration of acute illness and occurrence of post-leptospirosis symptoms of patients in the Netherlands. PLoS One. 2013;**8**(10):e76549. DOI: 10.1371/journal.pone.0076549

[65] Kurilung A, Chanchaithong P, Lugsomya K, Niyomtham W, Wuthiekanun V, Prapasarakul N. Molecular detection and isolation of pathogenic *Leptospira* from asymptomatic humans, domestic animals and water sources in Nan province, a rural area of Thailand. Research in Veterinary Science. 2017;**115**:146-154. DOI: 10.1016/j.rvsc.2017.03.017. Available from: https://www.sciencedirect.com/science/article/pii/S0034528816305380

[66] Moschini J. Leptospirosis humana con compromiso del sistema nervioso central. Comunicación de dos casos y revisión de la literatura. Neurología Argentina. 2011;**3**(4):222-228. Available from: https://www.elsevier.es/es-revista-neurologia-argentina-301-pdf-S1853002811000413

[67] Wilson MR, Naccache SN, Samayoa E, Biagtan M, Bashir H, Yu G, et al. Actionable diagnosis of neuroleptospirosis by next-generation sequencing. The New England Journal of Medicine. 2014;**370**(25):2408-2417. DOI: 10.1056/NEJMoa1401268

[68] Shpilberg O, Shaked Y, MaierMK SD, Samra Y. Long-term follow-up after leptospirosis. Southern Medical Journal. 1990;**83**(4):405-407. DOI: 10.1097/00007611-199004000-00011. Available from: https://pubmed.ncbi.nlm.nih.gov/2321070/

[69] Tenorio G, Rivas-Sánchez B, Velasco-Castrejón O. Utilidad de la determinación serológica de anticuerpos en el diagnóstico diferencial de la toxoplasmosis ocular con la leptospirosis ocular. Revista Médica del Hospital General de México. 2002;**65**(3):135-143

[70] Rathinam SR, Namperumalsamy P, Cunningham ET Jr. Spontaneous cataract absorption in patients with leptospiral uveitis. The British Journal of Ophthalmology. 2000;**84**(10):1135-1141. DOI: 10.1136/bjo.84.10.1135. Available from: https://www.researchgate.net/publication/12319671_Spontaneous_cataract_absorption_in_patients_with_leptospiral_uveitis

[71] Rathinam SR, Rathakrishnan S. Rapid maturation of unilateral cataract in leptospirosis. Indian Journal of Ophthalmology. 2020;**68**(9):1977-1979. DOI: 10.4103/ijo.IJO53520. Available from: https://www.ncbi.nlm.nih.gov/pmc/articles/PMC7690476/pdf/IJO-68-1977.pdf

[72] Pushpakumara J, Prasath T, Samarajiwa G, Priyadarshani S, Perera N, Indrakumar J. Myocarditis causing severe heart failure – An unusual early manifestation of leptospirosis: A case report. BMC Research Notes. 2015;**8**:80. DOI: 10.1186/s13104-015-1031-1

[73] Mitrakrishnan RN, Senaka R. Cardiac involvement in leptospirosis. Transactions of the Royal Society of Tropical Medicine and Hygiene. 2012;**106**:515-520. DOI: 10.1016/j.trstmh.2012.06.007

[74] Vaidya SR, Aeddula NR. Chronic kidney disease. In: StatPearls [Internet]. Treasure Island (FL): StatPearls Publishing; 2022. Available from: https://www.ncbi.nlm.nih.gov/books/NBK535404

[75] Kovesdy CP. Epidemiology of chronic kidney disease: An update 2022. Kidney International. Supplement. 2022;**12**(1):7-11. DOI: 10.1016/j.kisu.2021.11.003. Available from: https://www.ncbi.nlm.nih.gov/pmc/articles/PMC9073222/

[76] De Francesco DE, de Soares Abreu KL, da Bezerra SG Jr. Leptospirosis-associated acute kidney injury. Jornal Brasileiro de Nefrologia. 2010;**32**(4):400-407. English, Portuguese

[77] Ashford DA, Kaiser RM, Spiegel RA, Perkins BA, Weyant RS, Bragg SL, et al. Asymptomatic infection and risk factors for leptospirosis in Nicaragua. The American Journal of Tropical Medicine and Hygiene. 2000;**63**(5-6):249-254. Available from: https://pubmed.ncbi.nlm.nih.gov/11421372/

[78] Damesh GCH, Dilukshi SY. Chronic kidney disease of uncertain etiology in Sri Lanka: Are leptospirosis and Hantaviral infection likely causes? Medical Hypotheses. 2016;**91**:16-19. DOI: 10.1016/j.mehy.2016.04.009

[79] Li-Fang C, Huang-Yu Y, Cheng-Chieh H, Ya-Chung T, Shen-Hsing H, Chih-Wei Y. Leptospirosis kidney disease: Evolution from acute to chronic kidney disease. Biomedical Journal. 2023;**46**(4):100595. DOI: 10.1016/j.bj.2023.100595. Available from: https://www.ncbi.nlm.nih.gov/pmc/articles/PMC10345244/pdf/main.pdf

[80] Herath NJ, Kularatne SA, Weerakoon KG, Wazil A, Subasinghe N, Ratnatunga NV. Long term outcome of acute kidney injury due to leptospirosis? A longitudinal study in Sri Lanka. BMC Research Notes. 2014;7:398. DOI: 10.1186/1756-0500-7-398. Available from: https://www.ncbi.nlm.nih.gov/pmc/articles/PMC4080986/pdf/1756-0500-7-398.pdf

[81] Huang-Yy Y, Cheng-Chieh H, Su-Hsun L, Yi-Gen G, Yung-Chang C, Yi-Ching K, et al. Overlooked risk for chronic kidney disease after leptospiral infection: A population-based survey and epidemiological cohort evidence. PLoS Neglected Tropical Diseases. 2015;**9**(10):e0004105. DOI: 10.1371/journal.pntd.0004105. Available from: https://www.ncbi.nlm.nih.gov/pmc/articles/PMC4599860/pdf/pntd.0004105.pdf

[82] Chih-Wei Y. Leptospirosis renal disease: Emerging culprit of chronic kidney disease unknown Etiology. Nephron. 2018;**138**(2):129-136. DOI: 10.1159/000480691. Available from: https://karger.com/nef/article-pdf/138/2/129/3167379/000480691.pdf

[83] Chih-Hsiang C, Wei-Chiao S, Su-Wei C, Cheng-Chia L, Pei-Chun F, Huang-Yy Y, et al. Long-term outcome of leptospirosis infection with acute kidney injury. Biomedicine. 2022;**10**(10):2338. DOI: 10.3390/biomedicines10102338. Available from: https://www.ncbi.nlm.nih.gov/pmc/articles/PMC9598535/pdf/biomedicines-10-02338.pdf

[84] Trevejo RT, Rigau-Pérez JG, Ashford DA, McClure EM, Jarquín-González C, Amador JJ, et al. Epidemic leptospirosis associated with pulmonary Hemorrhage-Nicaragua, 1995. The Journal of Infectious Diseases. 1998;**178**(5):1457-1463. DOI: 10.1086/314424. Available from: https://academic.oup.com/jid/article/178/5/1457/810680?login=false

[85] Salkade HP, Divate S, Deshpande JR, Kawishwar V, Chaturvedi R, Kandalkar BM, et al. A study of autopsy findings in 62 cases of leptospirosis in a Metropolitan City in India. Journal of Postgraduate Medicine. 2005;**51**(3):169-173 Available from: https://www.bioline.org.br/pdf?jp05068

[86] Huda AS. The clinical picture, creating diagnostic constructs, and causation. In: The Medical Model in Mental Health: An Explanation and Evaluation. Online ed. Oxford: Oxford Academic; 2019, 2019. DOI: 10.1093/med/9780198807254.003.0004

[87] Couto RPB, Moura M, Drawanz HD. Molecular and serological diagnostic of leptospirosis: A review (2014-2020). Research, Society and Development. 2022;**11**(2):e19511225471. DOI: 10.33448/rsd-v11i2.25471. Available from: https://rsdjournal.org/index.php/rsd/article/view/25471

[88] Waggoner JJ, Pinsky BA. Molecular diagnostics for human leptospirosis. Current Opinion in Infectious Diseases. 2016;**29**(5):440-445. DOI: 10.1097/QCO.0000000000000295. Available from: https://www.ncbi.nlm.nih.gov/pmc/articles/PMC5127924/pdf/nihms831013.pdf

[89] Podgoršek D, Ružić-Sabljić E, Logar M, Pavlovic A, Remec T, Baklan Z, et al. Evaluation of real-time PCR targeting the lipL32 gene for diagnosis of *Leptospira* infection. BMC Microbiology. 2020;**20**:59. DOI: 10.1186/s12866-020-01744-4

[90] Kobayashi Y. Human leptospirosis: Management and prognosis. Journal of Postgraduate Medicine. 2005;**51**(3):201-204. Available from: https://pubmed.ncbi.nlm.nih.gov/16333193/

[91] Oriol LFB. Tratamiento de la leptospirosis humana. Alternativa antibiótica. Archivos de Medicina. 2015;**11**(2):2. Available from: file:///C:/Users/Beatriz/AppData/Local/Temp/MicrosoftEdgeDownloads/1b7351be-4389-49b0-9c98-57bb3b68d3cc/Dialnet-TratamientoDeLaLeptospirosisHumanaAlternativaAntib-5139233.pdf

[92] Berron-Perez R, Chavez-Sanchez R, Estrada-Garcia I, Espinosa-Padilla S, Cortez-Gomez R, Serrano-Miranda E, et al. Indications, usage, and dosage of transfer factor. Revista Alergia Mexico. 2007;**54**(4):134-139. Available from: https://www.researchgate.net/publication/5554117_Indications_usage_and_dosage_of_the_transfer_factor

[93] Navarrete EJ, Moreno MM, Rivas-Sánchez B, Velasco-Castrejón O. Leptospirosis prevalence in a population of Yucatán. Journal of Pathogens. 2011;**2011**:408604. DOI: 10.4061/2011/408604. Available from: https://pmc.ncbi.nlm.nih.gov/articles/PMC3335464/. Epub 2011 Nov 2

[94] Navarrete-Espinosa J, Rivas-Sánchez B, Soriano RML. Leptospirosis prevalence and associated factors in school children from Valle de Chalco-Solidaridad, state of Mexico. International Journal of Pediatric Research. 2015;**1**:2. DOI: 10.23937/2469-5769/1510008. Available from: https://www.researchgate.net/publication/299482268_Leptospirosis_Prevalence_and_Associated_Factors_in_School_Children_from_Valle_de_Chalco-Solidaridad_State_of_Mexico. ISSN: 2469-5769

Chapter 4

# Leptospirosis and Risk Factors: A Review

*Marie-Françoise Doursout, Claudia Ferreira, Joselito S. Balingit, Pooja Shivshankar and Yann A. Meunier*

## Abstract

Leptospirosis is a zoonotic disease with epidemic potential, especially after heavy rainfall, caused by a bacterium called *Leptospira, affecting humans and animals*. Leptospirosis occurs worldwide, but its incidence is the highest in tropical regions. Estimates indicate that there are more than 500,000 cases of leptospirosis each year worldwide. Several stains of *Leptospira* can induce an infection state primarily in various animal hosts such as rodents, cattle, and domestic animals, though human beings operate as secondary hosts. Therefore, Leptospirosis exhibits a broad diversity of clinical symptoms, beginning as a modest sickness and progressing to a more critical and deadly disease. Leptospirosis's symptoms mirror several disorders, such as dengue, influenza, as well as various viral hemorrhagic diseases. Consequently, a precise diagnosis both medical and biological tests is required when symptoms first occurred, primarily in outbreak situations. As such, in the present review, we describe the mode of transmission, specific symptoms and examples of leptospirosis outbreaks around the world. Despite all the preventive measures, leptospirosis remains a significant public health concern. Accordingly, we have highlighted the need to develop effective vaccines against leptospirosis for humans and animals.

**Keywords:** leptospirosis, pathogens, tropical diseases, environmental zoonoses, risk factors

## 1. Introduction

In the past decade, leptospirosis, also called Weill's disease or Nanukayami's disease (other names include canicola fever, canefield fever, and seven-day fever), come into sight as a worldwide crucial infectious disease. The disease arises mainly in urban industrialized and developed countries, as well as in rural provinces around the world (**Figure 1**). As such, mortality significantly persists, linked to gaps in adequate analysis and/or poor judgment, mainly due to poor groundwork and lack of resources. Likewise, other poorly understood reasons may include essential pathogenicity of some leptospiral strains. Leptospirosis, caused by bacterial pathogenic *Leptospira,* is an emerging worldwide zoonotic disease partially due to global warming and more frequent and severe flooding. Leptospirosis is also a waterborne disease (**Table 1**).



IntechOpen

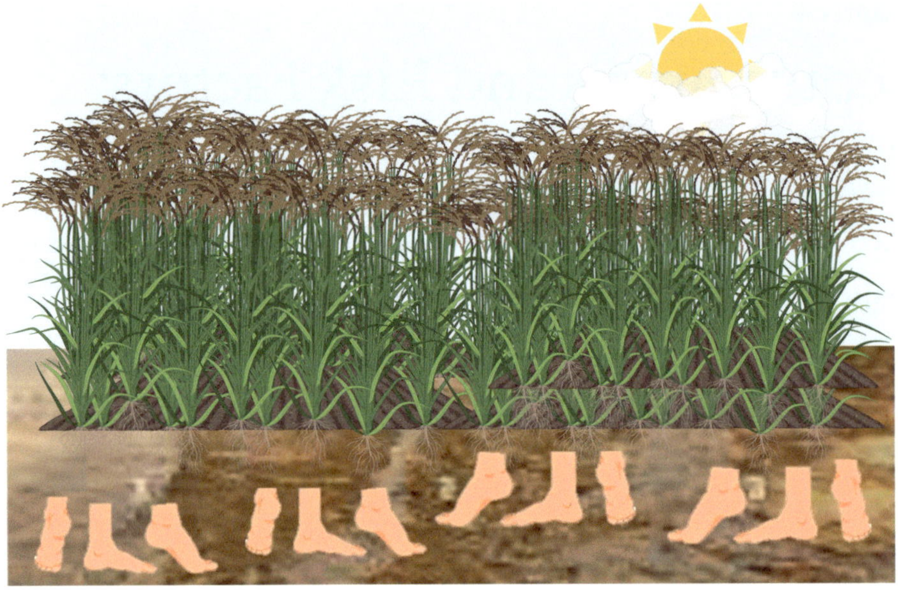

**Figure 1.**
*Working in a paddy field barefoot is a risk factor for leptospirosis. Source: This image is in the public domain.*

| |
|---|
| **Infectious agent: Serovars of** *Leptospira interrogans* |
| **Endemicity** |
| Global, higher incidence in humid and hot areas |
| **People at greatest risks to get infected** |
| Tourists, open-air athletes, as well as those exposed to fresh water or muddy terrain |
| Aid workers, mainly operating at sites of hurricanes or at various flooded regions |
| Military staff<br>Sewage professionals |
| **Preventive approaches** |
| Keep away from rat's urine, water or soil infected with animal's urine |
| Personal protective/shielding equipment |
| Chemoprophylaxis – Immunization |
| **Testing - diagnosis** |
| • Done by a clinical laboratory, certified in intricacy testing,<br>• Or State Health Department,<br>• Or connect to bspb@cdc.gov, the CDC's Bacterial Special Pathogens Branch for further specific diagnosis. |

**Table 1.**
*Infectious agent—Leptospira.*

Leptospires are slim and defined as helical bacteria (**Figure 2**). They are classified into 12 pathogenic and 4 saprophytic species, representing more than 250 pathogenic serovars [1, 2].

The primary reservoirs of Leptospira are rodents, serving as the main hosts [3]. When infected, rodents release these bacteria into the environment through their

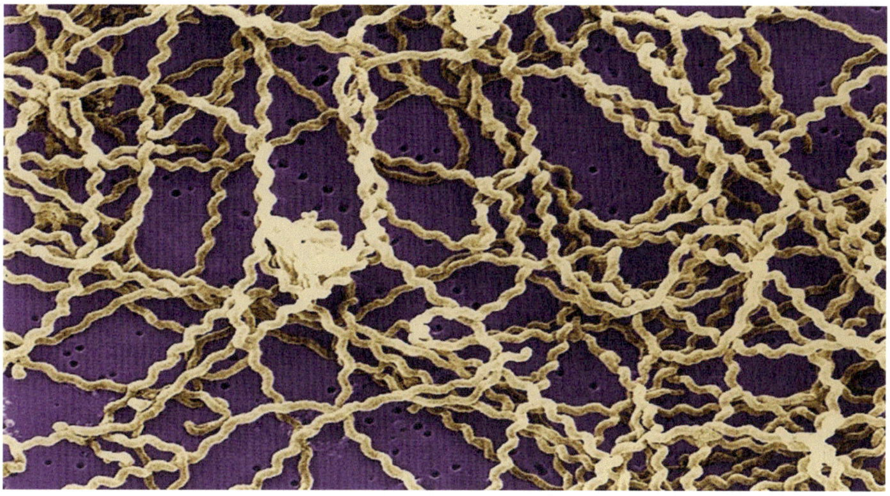

**Figure 2.**
*This digitally-colorized, scanning electron microscopic (SEM) image depicts a number of corkscrew-shaped,* Leptospira *sp. bacteria atop a 0.1 µm polycarbonate filter. Source: This image is in the public domain.*

urine, leading to contamination. Additionally, other animals such as dogs and livestock act as reservoirs, spreading pathogenic leptospires. As compared to rodents, these animals display severe clinical manifestations, involving abortion. As soon as the soil and water are infected with Leptospira, humans enter directly in contract through either an infected rats's blood or urine. The bacteria penetrates the human body mainly through lacerations of the skin, as well as the eyes and mouth mucosa. Following penetration, the bacteria enter the bloodstream and disseminate to various organs, causing pathological damage. However, interactions between environment, rodent reservoir and human leptospirosis remain poorly studied. Instances of Leptospirosis disease have been related following water consumption. Causes of infection may occur during inadequate water handling methods and/or deficient management.

## 2. Geographic distribution

One million severe cases are reported every year worldwide. Leptospirosis strikes around the world, yet in higher numbers in hot and humid locations [4]. Moreover, developed countries, have also perceived a significant increase in the number of reported leptospirosis cases (*European Centre for Disease Prevention and Control. Introduction to the Annual Epidemiological Report. 2020*). Available from: *https://www.ecdc.europa.eu/en/surveillance-and-disease-data/annual-epidemiologicalreports/introduction-annual*. Leptospirosis can be defined as a work-related disease affecting farmers, veterinarians and sewer workers or as a recreational disease related to water-bathing, canyoning and kayaking. Indeed, water exposure emerges as a major risk factor. Likewise, increases in seasonal floods amplified the risk of infection by exposing humans and animals to *Leptospira* that are washed out of their environment.

According to current literature, leptospira have a high incidence on public health in endemic regions and/or during specific seasons. However, little is known about the environmental concentration in urbanized areas, marked by the proliferation of rats, representing the leading animal reservoir of *Leptospira* [5]. Some explanations can be related to social exchanges and water-related activities that amplified in industrialized and developed countries, with the setting of urban beaches, bathing areas with non-treated surface water. As such, we have summarized few examples of leptospirosis infection in Europe and in the Pacific region in this review.

For instance, in France, a study led from 1995 to 2005 reported that 42% of patients became infected after performing water sports, 19% following contact with water from ponds, swamps, wells, water holes, and 19% during certified activities. [6]. It is well known that recreational water activities are becoming more and more prevalent. A recent study from Yvonnick Guillois described a gathering of 14 kayakers in Brittany, France, exhibiting leptospirosis symptoms following contact with water [7]. Likewise, Morgan et al. [8] examined an outbreak of leptospirosis in athletes and community locals after a triathlon held in Springfield, Illinois, USA, in 1998. A telephone survey was initiated to collect clinical information and data on possible risk factors. Likewise, community surveillance was created, animals and lake water samples were gathered to assess the source of leptospiral contamination. Data have shown that a total of 834 of 876 triathletes were contacted; 98 (12%) reported being sick. In addition, serum samples obtained from 474 athletes were tested; 52 of these samples (11%) tested positive for leptospirosis. Fourteen (6%) of 248 symptomatic community residents tested positive for leptospirosis. The authors concluded that heavy rains prior to the triathlon likely increased leptospiral contamination in Lake Springfield. As such, ingestion of one or more mouthfuls of lake water was a predominant risk factor for illness in athletes. This was the largest outbreak of leptospirosis reported in the United States [8].

A high incidence of leptospirosis was also reported in tropical areas, later named "the neglected disease" in the Pacific region, such as French Polynesia (FP), which is overseeing territory located in the South Pacific. Though data on the epidemiology in FP are rare, Bouscaren et al. [9] have described the likelihoods and surveillance of this disease to better focus on preventive methods. The authors have reported 11 years of surveillance data between January 1, 2007, and December 31, 2017. Over the study period, they have testified 1356 confirmed and likely reported leptospirosis cases. It was concluded that the mean annual incidence rate was 46.0 (95% confidence interval, 43.6–48.5) cases per 100. 000 residents. The investigators have registered 864 (63.7%) hospitalizations; of these, at least 270 (19.9%) were admitted in ICU (intensive care unit). Twenty-four patients (1.8%) died. The authors concluded that even if the occurrence of leptospirosis was minor in FP compared to most of other Pacific countries and territories, data confirmed that the disease is highly endemic in FP [9]. According to "la Depeche de Tahiti - https://ladepeche.pf," 8 cases of leptospirosis were identified in Tahiti, Raiatea, and Huahine on January 11, 2024. These cases were spotted following heavy rainfall or flooding in the regions. Among the infected cases, 5 individuals were hospitalized (62.5%). La Depeche de Tahiti reported that among 117 confirmed in 2023, 64% were males, age 15–65 from Tahiti (60%), Raiatea (22%), and Huahine (6%). Two deaths were recorded, both infected with leptospirosis [10].

Tubiana et al. [11] directed a retrospective case-control study of hospitalized patients confirmed with leptospirosis, admitted to two public hospitals in New Caledonia in 2008–2011. Patients exhibited fatal or severe leptospirosis, determined by clinical standards. Controls were defined as patients hospitalized for milder leptospirosis. Among the 176 patients enrolled in the study, 71 had criteria of severity including 10 deaths. The authors concluded that the study illustrated the benefit for clinicians to identify (i) the strain causing the infection, (ii) a life-threatening level of leptospiremia determined by qPCR and (iii) delivering results as soon as possible. Overall, Tubiana et al. indicated that preventive measures should include initial antibacterial therapy as well as rodent's control, specifically for the Ictero-Haemorrhagiae serogroup [11].

Furthermore, health experts have documented high level of leptospirosis activity throughout the Reunion Island including a total of 40 cases related from January 1 to February 13 2024, compared to only 13 cases documented during a similar time in 2023. Most of the cases were acknowledged in the South of the island (29 events), then West of the island with 5 events, East of the island with 4 events, and finally North of the island with solely 2 events. Health representatives widely advise inhabitants to consent on available disease preventions, such as wearing proper shielding clothes and equipment when working in fields, gardening and/or bathing activities, mainly in muddy terrain [12].

## 3. Symptoms

Leptospirosis can take different forms, from a flu-like disorder (myalgia, fever or headache) to a multisystem syndrome, with icteric and hemorrhagic pattern for 20% of cases. Leptospirosis symptoms frequently appear one to two weeks after infection. However, the incubation period can be as long as one month. The progression of leptospirosis disease can be divided into 2 phases: the early or febrile phase, characterized by modest flu-like symptoms, and the late or immune phase, scored by more severe clinical manifestations such as pulmonary diffuse hemorrhage and renal failure (**Figure 3**).

### 3.1 Early or febrile phase

Symptoms of the first phase last five to seven days. Typical clinical signs include conjunctival suffusion and a transient petechial outbreak. This transient skin outbreak is seen in most cases of leptospirosis and is identical to those observed

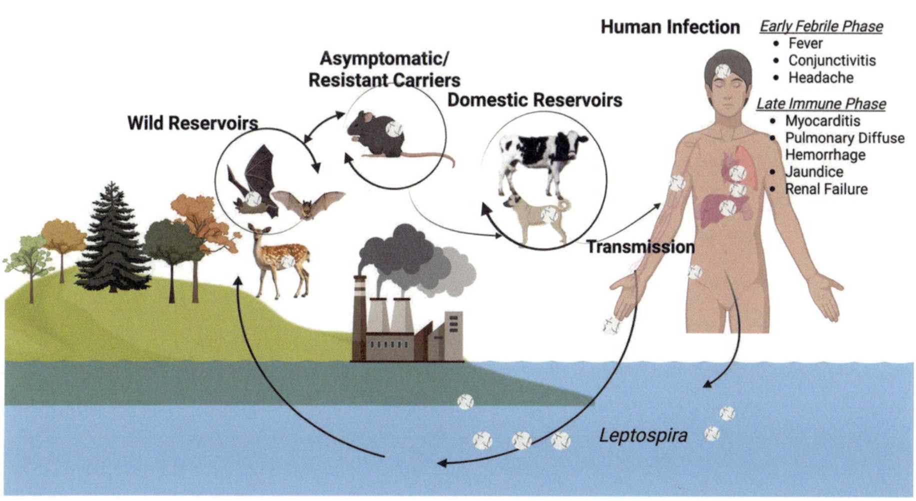


**Figure 3.**
*Cycle of Leptospirosis. This figure was generated using Bio-Render.*

in bacterial meningitis. Red, irregular spots appear on the skin, dark red in color, sometimes turning purple [13] (**Figure 4**).

### 3.2 Late or immune phase

In this phase, also called immune phase, symptoms settle as antibodies are produced against the bacteria. However, additional warning signs might develop. For instance, illness may not be well-defined, specifically in patients with harsh sickness. It has been reported that 90% of infected individuals undergo soft symptoms compared to 10% that undergo a more severe leptospirosis disease.

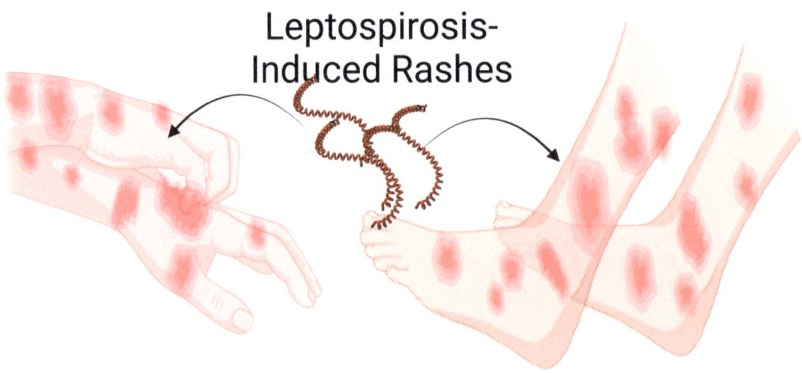


**Figure 4.**
*Leptospirosis—induced rashes. This figure was generated using Bio-Render.*

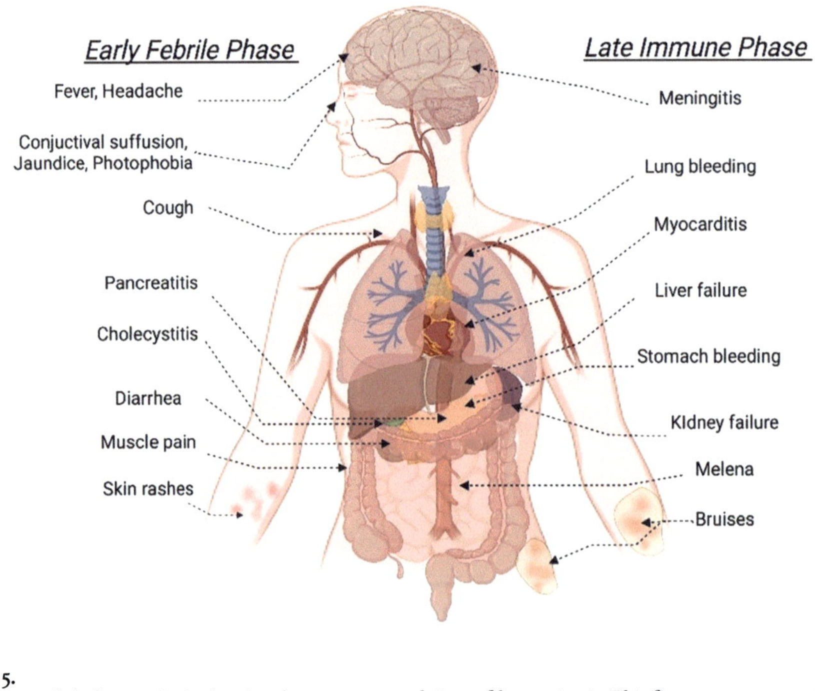


**Figure 5.**
*A schematic of the human body showing the symptoms and signs of leptospirosis. This figure was generated using Bio-Render.*

Severe symptoms may be accounted in the second phase or immune phase. This phase is distinguished by liver damage, causing jaundice, kidney failure, and bleeding, happening in 5–10% of those infected. Moreover, lung and brain damage can also arise. A variety of neurological complications such as partial paralysis of the body, complete inflammation of the horizontal section of spinal cord, and muscle softness due to nerve's damage can be observed in the classic form of leptospirosis. Signs of bleeding including non-traumatic contusions, nose bleeding, bloody stools due to bleeding in the stomach, vomiting blood and bleeding from the lungs can also be observed. Furthermore, a prolongation of prothrombin time is also observed following coagulation testing, indicating a severe bleeding symptom. However, low platelet count is not associated with severe bleeding. Pulmonary hemorrhage leads to massive coughing associated with blood, causing ARDS (acute respiratory distress syndrome), with a risk of death at 50% [14–16]. Complications of the disease can occur as myocarditis, aseptic meningitis, hepatic failure with or without renal failure, pulmonary hemorrhage, iridocyclitis, seizures, ARDS, and coma [17]. Symptoms occurring in humans in both phases are summarized in (**Figure 5**).

## 4. Risk factors

Risk factors can be divided into 2 categories (summarized in **Table 2**)

1. individual categories

2. collective / shared categories

## 5. Provision care

In the event of strong sickness, Wiwanitkit et al. [18] reported that provision care including renal substitute treatment, ventilation maintenance, and blood samples

| |
|---|
| *Individual categories* |
| Contact with contaminated stagnant water or mud |
| Contact with infected rodents directly or indirectly through food materials they consume |
| Leisure: Hunting, swimming in contaminated water (triathlon, open water swimming); sailing/windsurfing/rafting/golfing); freshwater fishing. |
| Venture tourist/explorer |
| Skin wound or cut |
| *Collective / shared categories* |
| Hurricanes, heavy rain, flooding |
| Poor sewage infrastructure |
| Poor rat population control |
| Animal urine in the soil and other surfaces |
| Sewer worker; Fish farm worker; Street dweller/urban slums |

**Table 2.**
*Risk factors.*

analysis may also be required [18]. Such therapy mimics organ failure, seen for instance in septic patients. In a Brazilian study performed in patients with leptospirosis complications associated with acute respiratory distress syndrome (ARDS) and acute kidney injury (AKI), hemodialysis administered daily was reported valuable as compared to alternative dialysis [19]. Abdulkader et al. [20] reported that hypokalemia is common in non-oliguric AKI associated with leptospirosis and should be corrected [20].

## 6. Therapy

It is well established that most leptospirosis events are self-limited without antimicrobial treatment. However, some individuals build up clinical problems resulting in elevated deaths. Overall, in the event the disease is strong enough to benefit clinical care and the establishment of a diagnosis, antibiotic therapy should be given to reduce the length of the illness. The approach varies with the clinical presentation. A Jarisch-Herxheimer reaction may arise after leptospirosis's antimicrobial therapy. This reaction is described as acute inflammatory response in response to the clearance of spirochetes from the blood, clinically described by fever, muscle rigors, and decrease in blood pressure [21–23]. This inflammatory response occurred in 21% of cases from a total of 262 patients with leptospirosis. For outpatients with mild disease, doxycycline or azithromycin treatments are in favor. These agents also have activity against rickettsial disease, which can be confused with leptospirosis. Azithromycin is preferred over amoxicillin if the differential diagnosis includes rickettsial infection. For hospitalized adults with severe disease, Nick Day et al. [24] favor treatment with penicillin, doxycycline and ceftriaxone (1 to 2 g, IV, q24h), or cefotaxime (1 g, IV, q6h). The duration of treatment in severe disease is usually 7 days [24].

Given the vasculitic nature of severe leptospirosis, particularly in the setting of pulmonary participation, corticosteroid therapy, administered intravenously has been suggested. Currently, little is known for corticosteroids usage. However, some reports have proposed a beneficial effect in using steroids in addition to antibiotic therapy in rigorous illness [25–29].

## 7. Prevention

Although all protective measures have been deployed, leptospirosis disease remains a strong medical concern worldwide. Therefore, the need to continue rigorous leptospirosis observation, training medical staff, providing material to the public, remain to be emphasized [30]. Anticipation procedures comprise escaping latent sources of infection, to administer prophylaxis for people with high risk of contact, as well as preceding to vaccinate all animals. Specifically, the most crucial control measures to anticipate human leptospirosis include: refrain from standing water, rodent restriction, and finally avoiding consumable food from contaminated animals. Antimicrobial prophylaxis for people at high risk of contact may be useful in some situations. However, evidence is vulnerable as investigations have been mixed in experimental design and mainly underpowered [31]. A study conducted by Sehgal et al. [32], enrolled 700 people in the Andaman Islands, an extremely endemic location in Southeast Asia. The authors reported that the occurrences of leptospirosis are secondary to high water, leading to flooding. Individuals were randomized to doxycycline 200 mg, orally administered weekly, or placebo. Results demonstrated that the rates of clinical infection were attenuated in individuals receiving doxycycline (3.1 vs.6.8%), although no changes in the rates of seroconversion were reported [32]. In a different study conducted by Takafuji et al. [33] 900 warfighters positioned in the Panama jungle for exercises and drilling were enrolled. The investigators reported that a small amount of leptospirosis cases were detected among individuals who took doxycycline treatment (200 mg orally administered every week for 2 to 3 weeks and at the end of exposure) compared with placebo (1 vs. 20 cases) [33].

## 8. Vaccination

According to the CDC, there is no vaccine approved for use in the United States to prevent leptospirosis in people, though some countries have approved. Existing anti-leptospirosis vaccines in the market consist of whole inactivated Leptospira, commonly referred to as bacterin. Up to now, only bacterins have obtained approval for human application. Bacterins consist of one (monovalent) or more (polyvalent) chemically inactivated pathogenic Leptospira serovars, achieved through methods like ethanol, formaldehyde, phenol, or physical approaches e. g. heat, irradiation, or freezing.

Immunization of domestic and farm animals against leptospirosis can provide variable levels of protection [34–37]. Some immunized animals become infected and excrete leptospires in their urine [38].

Vaccination to combat leptospirosis in humans, has failed in clinical treatment despite a significant number of candidates willing to be tested over years [39, 40]. Interestingly, commercial vaccines to treat human leptospirosis are solely available in Japan, China, Cuba and France. Isabela Resende Azevedo et al. [41] reported that commercial vaccines primed with inactivated Leptospira (bacterins) generate a short and serovar-specific immune response, associated with robust adverse consequences [41].

## 9. Conclusion

In this extensive review, we discuss the global challenge posed by the rise of leptospirosis due to climate change and increase in rodent populations. We conclude that leptospirosis has emerged as a significant global health concern over the past decade, affecting both urban and rural populations worldwide. The disease, caused by pathogenic Leptospira bacteria, presents challenges in diagnosis, treatment, and prevention due to factors such as inadequate infrastructure, environmental changes, and a lack of effective vaccines.

The transmission of leptospirosis primarily occurs through contact with water or soil contaminated with the urine of infected wild animals, particularly rodents. Interestingly, it has been reported that many animals in the wild including mice, rats, squirrels, raccoons, opossums, skunks, deers and foxes transport the bacteria without developing the disease [42]. Early diagnosis and appropriate antibiotic therapy are essential for improving clinical outcomes, especially in severe cases.

Preventive measures play a crucial role in controlling leptospirosis, including avoiding exposure to contaminated water, implementing rodent control measures, and providing prophylactic antibiotics to high-risk individuals. However, despite these efforts, leptospirosis remains a persistent public health challenge, particularly in tropical regions prone to flooding and without adequate sanitation infrastructure.

Furthermore, the development of effective vaccines against leptospirosis for both humans and animals remain a priority. While existing vaccines offer limited protection and may have adverse side effects, ongoing research aims to improve vaccine efficacy and safety to mitigate the burden of this disease.

Finally, addressing leptospirosis requires a multifaceted approach involving enhanced surveillance, improved diagnostic capabilities, better infrastructure, and continued research into vaccines and therapeutic interventions. By implementing comprehensive strategies, we can strive better to reduce the global burden of leptospirosis and improve the health outcomes of affected populations.

## Author details

Marie-Françoise Doursout[1*], Claudia Ferreira[2], Joselito S. Balingit[3], Pooja Shivshankar[4] and Yann A. Meunier[5]

1 Department of Anesthesiology, Critical Care and Pain Medicine, UTHealth-McGovern Medical School, Houston, TX, USA

2 Sorbonne University, Paris, France

3 TECHNE Medical Department Champignelles, France

4 Center for Immunology and Autoimmune Diseases, Institute of Molecular Medicine, UTHealth-McGovern Medical School, Houston, TX, USA

5 International Institute of Medicine and Science, Rancho Mirage, CA, USA

*Address all correspondence to: marie-francoise.doursout@uth.tmc.edu

**IntechOpen**

© 2025 The Author(s). Licensee IntechOpen. This chapter is distributed under the terms of the Creative Commons Attribution License (http://creativecommons.org/licenses/by/3.0), which permits unrestricted use, distribution, and reproduction in any medium, provided the original work is properly cited.

## References

[1] Faine S, Adler B, Bolin C, Perolat P. Leptospira and Leptospirosis. 2nd ed. Melbourne, Australia: Medisci; 1999

[2] Haake DA, Levett PN. Leptospirosis in Humans. In: Adler B, editor. Leptospira and Leptospirosis. Current Topics in Microbiology and Immunology. Vol. 387. Berlin, Heidelberg: Springer-Verlag; 2015. pp. 65-97. DOI: 10.1007/978-3-662-45059-8_5. ISBN 978-3-662-45058-1

[3] Lau C, L, Townell N, Stephenson E, van den Berg D, Craig SB. Leptospirosis an important zoonosis acquired through work, play and travel. AJGP. 2018;**47**(3):105-110

[4] Bharti AR, Nally JE, Ricaldi JN, Matthias MA, Diaz MM, Lovett MA, et al. Peru-United States Leptospirosis Consortium. Leptospirosis: A zoonotic disease of global importance. The Lancet Infectious Diseases. 2003;**3**(12):757-771. DOI: 10.1016/s1473-3099(03)00830-2

[5] Richard E, Geslin J, Wurtzer S, Moulin L. Monitoring of Leptospira species diversity in freshwater bathing area and in rats in Paris, France. Science of the Total Environment. 2022;**833**:155121. DOI: 10.1016/j.scitotenv.2022.155121 Epub 2022 Apr 7. DOI:

[6] Strady C, Penalba C, Baumard S, et al. Nephropathia epidemica and leptospirosis in Champagne-Ardenne, France: Comparison of clinical, biological and epidemiological profiles. European Journal of Clinical Microbiology & Infectious Diseases. 2009;**28**:825-829. DOI: 10.1007/s10096-009-0716-6

[7] Guillois Y, Bourhy P, Ayral F, Pivette M, Decors A, Grau JHA, et al. An outbreak of leptospirosis among kayakers in Brittany, north-West France, 2016. Euro Surveillance. 2018;**23**(48):1700848. DOI: 10.2807/1560-7917.ES.2018.23.48.1700848

[8] Morgan J, Bornstein SL, Karpati AM, Bruce M, Bolin CA, Austin CC, et al. Clinical Infectious Diseases. 2002;**34**(12):1593-1599. DOI: 10.1086/340615. Epub 2002 May 24

[9] Bouscaren N, Coignac S, Lastère D, Musso Y, Teissier J, Formont FC, et al. Leptospirosis in French Polynesia: 11 years of surveillance data, 2007-2017. New Microbes and New Infections. 2019;**29**:100518

[10] La Depeche de Tahiti. Available from: https//:ladepeche.pf [Accessed: 2024 January 11]

[11] Tubiana S, Mikulski M, Becam J, et al. Risk factors and predictors of severe leptospirosis in New Caledonia. PLoS Neglected Tropical Diseases. 2013;**7**:e1991

[12] CDC. Preventing Leptospirosis after Hurricanes or Flooding. Available from: https://www.cdc.gov/leptospirosis/prevention/index.html

[13] Higny J, Forêt F, LaterreP-F. Leptospirosis-induced purpura: An atypicalmanifestation of Weil's disease. Clinical Case Reports. 2020;**8**:572-573. DOI: 10.1002/ccr3.2692

[14] Karpagam KB, Ganesh B. Leptospirosis: A neglected tropical zoonotic infection of public health importance-an updated review. European Journal of Clinical Microbiology & Infectious Diseases. 2020;**39**(5):835-846

[15] "Factsheet about leptospirosis" European Centre for Disease Prevention and Control (ECDPC). 2010

[16] Cagliero J, Villanueva SY, Matsui M. Leptospirosis pathophysiology: Into the storm of cytokines. Frontiers in Cellular and Infection Microbiology. 2018;**8**(204):204. DOI: 10.3389/fcimb.2018.00204

[17] Yann M. Tropical Diseases: A Practical Guide for Medical Practitioners and Students. Vol. 162. New York: Oxford University Press; 2013. Available from: https://www.amazon.com/Tropical-Diseases-Practical-Practitioners-Students-ebook/dp/B00FI001X8/ref=sr_1_4?qid=1660257416&refinements=p_27%3AYann+A.+Meunier&s=books&sr=1-4

[18] Wiwanitkit V. Comparison between blood exchange and classical therapy for acute renal failure in Weil's disease: Appraisal on Thai reports. Nephrology (Carlton). 2006;**11**:481

[19] Andrade L, Cleto S, Seguro AC. Door-to-dialysis time and daily hemodialysis in patients with leptospirosis: Impact on mortality. Clinical Journal of the American Society of Nephrology. 2007;**2**:739

[20] Abdulkader RC, Seguro AC, Malheiro PS, et al. Peculiar electrolytic and hormonal abnormalities in acute renal failure due to leptospirosis. The American Journal of Tropical Medicine and Hygiene. 1996;**54**:1

[21] Guerrier G, D'Ortenzio E. The Jarisch-Herxheimer reaction in leptospirosis: A systematic review. PLoS One. 2013;**8**:e59266

[22] Tunbridge AJ, Dockrell DH, Channer KS, McKendrick MW. A breathless triathlete. Lancet. 2002;**359**:130

[23] Takamizawa S, Gomi H, Shimizu Y, et al. Leptospirosis and Jarisch-Herxheimer reaction. QJM. 2015;**108**:967

[24] Day N. Leptospirosis: Treatment and prevention. https://www.uptodate.com/contents/leptospirosis-treatment-and-prevention

[25] Kularatne SA, Budagoda BD, de Alwis VK, et al. High efficacy of bolus methylprednisolone in severe leptospirosis: A descriptive study in Sri Lanka. Postgraduate Medical Journal. 2011;**87**:13

[26] Thunga G, John J, Sam KG, et al. Role of high-dose corticosteroid for the treatment of leptospirosis-induced pulmonary hemorrhage. Journal of Clinical Pharmacology. 2012;**52**:114

[27] Minor K, Mohan A. Severe leptospirosis: Treatment with intravenous corticosteroids and supportive care. The American Journal of Emergency Medicine. 2013;**31**:449.e1

[28] Azevedo AF, Miranda-Filho Dde B, Henriques-Filho GT, et al. Randomized controlled trial of pulse methyl prednisolone × placebo in treatment of pulmonary involvement associated with severe leptospirosis. [ISRCTN74625030]. BMC Infectious Diseases. 2011;**11**:186

[29] Rodrigo C, Lakshitha de Silva N, Goonaratne R, et al. High dose corticosteroids in severe leptospirosis: A systematic review. Transactions of the Royal Society of Tropical Medicine and Hygiene. 2014;**108**:743

[30] Goarant C. Leptospirosis: Risk factors and management challenges in developing countries. Research and Reports in Tropical Medicine. 2016;7:49-62. Published online 2016 Sep 28. DOI: 10.2147/RRTM.S102543

[31] Guzmán, Pérez M, Blanch Sancho JJ, Segura Luque JC, et al.

Current evidence on the antimicrobial treatment and chemoprophylaxis of human leptospirosis: A meta-analysis. Pathogens. 2021;**10**(9):1125

[32] Sehgal SC, Sugunan AP, Murhekar MV, et al. Randomized controlled trial of doxycycline prophylaxis against leptospirosis in an endemic area. International Journal of Antimicrobial Agents. 2000;**13**:249

[33] Takafuji ET, Kirkpatrick JW, Miller RN, et al. An efficacy trial of doxycycline chemoprophylaxis against leptospirosis. The New England Journal of Medicine. 1984;**310**:497

[34] André-Fontaine G, Branger C, Gray AW, Klaasen HL. Comparison of the efficacy of three commercial bacterins in preventing canine leptospirosis. The Veterinary Record. 2003;**153**:165

[35] Hancock GA, Wilks CR, Kotiw M, Allen JD. The long term efficacy of a hardjo-Pomona vaccine in preventing leptospiruria in cattle exposed to natural challenge with Leptospira interrogans serovar hardjo. Australian Veterinary Journal. 1984;**61**:54

[36] Rinehart CL, Zimmerman AD, Buterbaugh RE, et al. Efficacy of vaccination of cattle with the Leptospira interrogans serovar hardjo type hardjoprajitno component of a pentavalent Leptospira bacterin against experimental challenge with Leptospira borgpetersenii serovar hardjo type hardjo-bovis. American Journal of Veterinary Research. 2012;**73**:735

[37] Sykes JE, Hartmann K, Lunn KF, et al. 2010 ACVIM small animal consensus statement on leptospirosis: Diagnosis, epidemiology, treatment, and prevention. Journal of Veterinary Internal Medicine. 2011;**25**:1

[38] Gomes-Solecki M, Santecchia I, Werts C. Animal models of leptospirosis: Of mice and hamsters. Frontiers in Immunology. 2017;**8**:58. DOI: 10.3389/fimmu.2017.00058.eCollection.2017

[39] Barazzone GC et al. Revisiting the development of vaccines against pathogenic Leptospira: Innovative approaches, present challenges, and future perspectives. Frontiers in Immunology. 2022, 2021;**12**:1-11. DOI: 10.3389/fimmu.2021.760291

[40] Lane AB, Dore MM. Leptospirosis: A clinical review of evidence based diagnosis, treatment and prevention. World Journal of Clinical Infectious Diseases. 2016;**6**(4):61. DOI: 10.5495/wjcid.v6.i4.61. ISSN 2220-3176

[41] Azevedo IR, Amamura TA, Isaac L. Human leptospirosis: In search for a better vaccine. Scandinavian Journal of Immunology. 2023;**98**(5):1-21. DOI: 10.1111/sji.13316

[42] Levett PN. Leptospirosis. Clinical Microbiology Reviews. 2001;**14**:296-326. DOI: 10.1128/cmr.14.2.296-326.2001

Chapter 5

# Leptospirosis: A Complex Disease

*Durlav Prasad Bora, Chandrani Goswami, Sophia Makdoh Gogoi, Arfan Ali, Nur Abdul Kader, Kalimuthusamy Natarajaseenivasan and Nagendra Nath Barman*

## Abstract

Leptospirosis, a major issue in veterinary medicine and public health, is a bacterial systemic disease caused by pathogenic spirochetes of the genus *Leptospira*. This infectious disease affects a wide range of animal species, including humans, indicating its widespread impact. Leptospires are highly diverse, with over 20 species, which are divided into over 300 serovars and more than 24 serogroups, which complicates the pathogenicity of the disease. Leptospirosis causes economic losses in livestock, which has larger consequences for food security and farming communities. Leptospirosis develops differently in livestock and wildlife, with small mammals, especially rodents, playing a crucial role as maintenance hosts. Effective management requires surveillance systems that employ serological, molecular, and cultural approaches. The interdependence of human and animal health emphasizes the need for a worldwide One Health approach that prioritizes collaborative work, research, and targeted interventions to address the multifaceted challenges of leptospirosis in a unified manner.

**Keywords:** leptospirosis, animal, zoonoses, epidemiology, diagnosis & Control

## 1. Introduction

Leptospirosis is a bacterial systemic infection caused by pathogenic spirochetes of the genus *Leptospira* which affects domestic animals and wildlife, as well as humans [1, 2]. Leptospirosis presents an important public health concern, occurring at the convergence of the animal-human-environment ecosystem. Geographic and host factors impact its complexities, making it a challenging issue [3]. It is a neglected re-emerging zoonosis and it is one of the six diseases identified by the World Organization for Animal Health (WOAH) in relation to animal diseases and climate change [1, 4]. The disease presents itself as a much greater incidence in tropical regions as well as in developing economy [1, 2, 5, 6]. In 1886, Weil defined leptospirosis as acute jaundice; Weil's disease was characterized by renal failure and extensive hemorrhage [7, 8]. However, Weil did not identify or isolate the causal agent. Several decades later, in 1914, Inada and Ido became the first scientists to detect leptospires after inoculating a guinea pig liver with the blood of a jaundiced patient and observing a spirochete, which they named *Leptospira haemorrhagae*. A report was released in 1915 explaining their discovery and proposing this spirochete as the primary agent of Weil's disease [9]. Leptospirosis constitutes a significant economic challenge, especially when it influences the reproductive performance of food-producing animals [10–12]. This adverse effect extends to various aspects of reproduction, including abortion, embryonic death, stillbirths, and infertility, leading to significant economic losses due to lower meat and milk outputs [13, 14]. Furthermore, the disease contributes to mastitis, weak pregnancy, decreased milk production and growth rates, enforcing indirect treatment costs [15, 16]. These losses are most evident in cattle and pigs, which are usually considered to be more susceptible than small ruminants [17, 18].

## 2. Etiology

*Leptospira*, derives its name from the Greek 'leptos' meaning fine and Latin 'spira' indicating coil, represents ubiquitous spirochetes. Leptospirosis is caused by Gram-negative bacterium within the Spirochaetes phylum, Leptospiraceae family, and *Leptospira* genus [1, 2, 19]. These spiral-shaped microorganisms measure 0.1 µm in diameter and 6–20 µm in length, featuring a pointed end typically folded into a characteristic hook shape, giving them a distinctive question-mark shape [1, 19]. *Leptospira* exhibits high mobility, executing rotational movements, translation, undulation, and flexion facilitated by two periplasmic axial flagella located under the cell membrane [20]. Despite being a microaerophile, *Leptospira* thrives even in conditions of complete aerobiosis, unable to survive in dry environments, heat, acidic, and basic disinfectants. Its optimal growth temperature ranges between 28 and 30°C, with the ability to grow at 37°C, and the ideal pH falls between 7.2 and 7.4 [2].

## 3. Classification of Leptospiraceae

The genus *Leptospira* is one of the most varied genera of bacteria, with species identified all over the world. The first taxonomic classification was based on the serological similarities of the strains [21]. The genus *Leptospira* has been conventionally divided into two groups, saprophytes—*Leptospira biflexa sensu lato*—and pathogens—*Leptospira interrogans sensu lato*—based on their virulence. These species are further divided into numerous serovars determined by the agglutination reaction due to the surface lipopolysaccharide of the bacteria. There are over 20 species of *Leptospira*, which are divided into over 300 serovars and more than 24 serogroups [22–24]. A new classification system based on genetic classification reveals at least 19 species [25, 26], with seven, including *L. interrogans*, *L. borgpetersenii*, *L. santarosai*, *L. noguchii*, *L. weilli*, *L. kirschneri*, and *L. alexanderi*, being the primary agents of leptospirosis [26–29]. The development of genetic typing technologies has accelerated the identification of serovars, which may now be used as alternative typing systems to identify a group of serovars that cause leptospirosis. More recently, a phylogenetic study found that *Leptospira* may be classified into three groups based on pathogenicity status, namely, pathogenic, non-pathogenic, and intermediate groups [30].

## 4. Epidemiology

### 4.1 Animal hosts and reservoirs

Reservoirs are animals that stay substantially asymptomatic in spite of being infected. Reservoirs retain leptospires in their renal tubules, where they grow and reproduce, and periodically release them through urination [31]. Although rodents are the most common reservoirs [32, 33], other animals such as bats [34], small terrestrial mammals [35, 36], and cattle [37] have also been identified as major *Leptospira* reservoirs. The basic life cycle of pathogenic *Leptospira* includes wild or domestic animals acting as reservoirs, releasing bacteria into the environment *via* their urine. Feral and wild animals also act as reservoir hosts for leptospires and often do not exhibit clinical signs of the disease [38, 39], releasing bacteria into the environment through their urine and infecting humans and domestic animals, which then may become a source of infection for other animals and humans [39, 40].

A maintenance host is an animal infected with a host-adapted serovar of the pathogen. Although each serovar is adapted to a specific maintenance host, it can cause disease in any mammalian species [41]. Clinically, a serovar reacts differently in its maintenance host species and in incidental or accidental hosts [42]. Leptospires are maintained in nature by chronic infection of the renal tubules of maintenance hosts [19]. Examples include serovar Bratislava in swine and serovar hardjo and bovis in cattle [42]. These animals can shed leptospires in their urine for years, potentially transmitting the infection to additional susceptible reservoir hosts or accidental (incidental) hosts [43]. The principal reservoir animals for the majority of *Leptospira* serovars are wild animals, notably rodents. Accidental or incidental infection occurs when vulnerable animals are exposed to non-host-adapted serovars. The incidental host has a low sensitivity to infection but a high pathogenicity for the host, causing rapid and severe infection rather than chronic disease. Infection by the pig-adapted serovar Pomona in cattle is one such example [42]. Humans serve as accidental hosts for *Leptospira* species [44].

### 4.2 Source of infection and transmission routes

Contaminated water is a major source of *Leptospira* bacteria. Stagnant or slow-flowing water bodies create breeding grounds, and animals can get sick by drinking or coming into contact with water contaminated by the urine of diseased animals. Furthermore, the urine of carrier animals, particularly rats, acts as a main source [45, 46]. Leptospirosis spreads among animals primarily through direct contact, involving interactions with infected urine or proximity to carrier animals. In crowded living areas, this transmission mode is particularly relevant. Animals can also contract the disease by ingesting contaminated food or water (**Figure 1**). *Leptospira* enters the host through mucosa and broken skin, resulting in bacteremia [47]. Humans can contract the infection through direct contact with infected animals, their urine, or contaminated environments [48]. This highlights the interconnectedness of the health of animals and humans in the context of leptospirosis. Human activities, including agricultural practices and urbanization, can influence the prevalence of leptospirosis in animals and subsequently impact the risk of zoonotic transmission [49, 50].

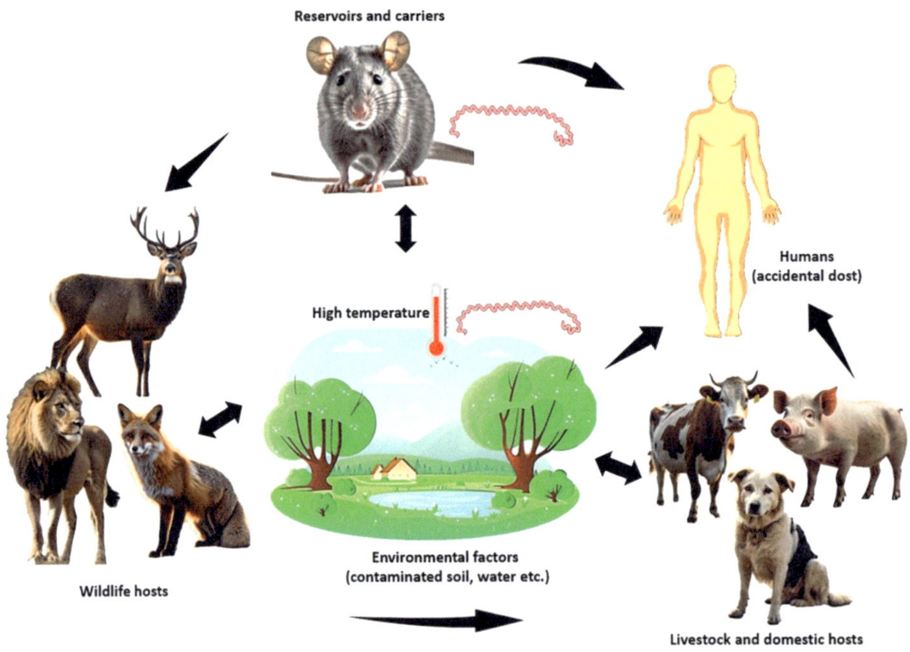


**Figure 1.**
*Transmission cycle of Leptospirosis.*

### 4.3 Seasonal patterns and climate influence

Seasonal variations and climate impact the dynamics of leptospirosis in animals, influencing disease prevalence, transmission, and emergence. The interplay of environmental conditions and the biology of *Leptospira* bacteria contributes to regionally diverse seasonal patterns. In warm and humid settings, *Leptospira* bacteria can survive better. Leptospires may survive in soil and water for long periods of time, especially when temperatures and moisture levels are favorable. As a result, tropical and subtropical climates frequently have increased leptospirosis prevalence [51]. There is substantial evidence that pathogenic *Leptospira* may survive and stay infectious in the environment for months, particularly in soils [52, 53]. According to other researchers, leptospires are more commonly isolated from soils than from freshwater samples [54–56]. These findings support the concept that soils might serve as a reservoir for *Leptospira* spp., especially pathogenic strains [57, 58]. Seasonal fluctuations during rainfall also have a considerable impact on the disease dynamics. Heavy rains and flooding events generate excellent conditions for the spread of leptospires [59]. Contaminated water sources can increase the risk of exposure for animals [60, 61]. Flooding can cause sick animals to be displaced, adding to the transmission cycle.

Climate phenomena such as El Niño and La Niña, particularly in the Equatorial Pacific, can have a global influence on rainfall patterns [62, 63]. These events may cause unexpected weather patterns, which may increase the prevalence of leptospirosis in locations prone to such climatic variations. The combination of excessive rainfall, warm temperatures, and high humidity provides an environment favorable to bacterial survival and transmission [64, 65]. This seasonality is most noticeable in locations with distinct wet and dry seasons. Warmer temperatures may raise the multiplication rates of *Leptospira* bacteria. This biological factor, together with seasonal temperature fluctuations, leads to the varied patterns of leptospirosis occurrence [66]. Water availability may become scarce in places with distinct dry seasons, causing reservoir hosts to congregate around surviving water bodies. This aggregation increases the likelihood of *Leptospira* transmission, as animals share limited water resources. Seasonal variations can also alter animal behavior and migration, influencing their interactions with contaminated surroundings [67]. During the wet season, for example, animals may seek higher ground or move to new habitats, altering the patterns of *Leptospira* exposure.

### 4.4 Global distribution

Leptospirosis has worldwide distribution caused by infection with any of several pathogenic serovars of *Leptospira*. There is a predominance of several *Leptospira* serovars (Icterohaemorrhagiae, Pomona, Canicola, Bratislava, Wolffi, Hardjo, and Grippotyphosa) in America, with a high concentration in the United States, Canada, Argentina, and Colombia [68]. Animal serovar variability was documented to be more prominent in Brazil and the United States [68]. In bovines, the most common recorded serovars were Bratislava, Icterohaemorrhagiae, and Hardjo in Brazil [69, 70]. Similarly, leptospirosis has been documented in practically all geographic zones of Africa [71]. The prevalent serovars, including Sokoine, are linked to cattle and rodents; Canicola, detected in rodents; and Grippotyphosa, observed in cattle [71]. The disease has also been seen in a variety of animal species in several European nations. The prevailing serovars could differ by region, and the disease is being identified in wild as well as domesticated animals [72–75]. The distribution of *Leptospira* serovars in animals in Asia is diverse, with a significant impact on animal agriculture. The most common serovars prevalent in India were Autumnalis, Grippotyphosa, Pyrogenes, Australis, Canicola, Javanica, Ballum, Sejroe, Louisiana, and Pomona among cattle [76–78]. Another research in India found that the most likely serovars to cause disease in dogs were Icterohaemorrhagiae, Pomona, Canicola, Autumnalis, and Grippotyphosa [79]. A study in Thailand recognized that the most commonly detected *Leptospira* serovars in cattle were Sejroe and Mini, whereas Mini, Sejroe, and Bratislava were mostly detected in buffaloes, Ranarum, Pomona, and Bratislava in pigs, and Mini, Shermani, and Ranarum in sheep and goats [80]. Research in Malaysia found a significant enzootic incidence in the animal population. Leptospirosis is considered endemic in Vietnam, Cambodia, and Laos [81]. Over 19 leptospiral serovars from over 7 serogroups have also been found in the Asia Pacific region [77].

## 5. Clinical manifestations in animals

Leptospirosis has a wide range of clinical manifestations in various animal species, reflecting the broad host range of this zoonotic pathogen. Leptospirosis manifests clinically similar in cattle and buffalo, posing considerable economic implications. In cattle and buffalo, clinical signs of leptospirosis include infertility, abortion, low milk output, and reproductive issues [76, 77, 82–84]. The disease can cause significant economic losses in the cattle sector owing to lower milk output and reproductive difficulties [84]. In piglets, particularly those in their early stages of development, signs of leptospirosis may include fever, lack of appetite, hemolytic anemia, hemoglobinuria, icterus, and infrequent convulsions. Some piglets may fail to develop and gain weight as expected [16, 42]. The signs of chronic leptospiral infection in dams often appear as diverse reproductive failures, resulting in low conception rates. Infected pregnant animals with the infection often experience late-term abortions [16, 19]. In dogs, leptospirosis can cause a number of clinical manifestations, including renal disease, hepatic disease, uveitis, and pulmonary hemorrhage The severity of the condition varies, with some dogs displaying no or few symptoms while others suffer serious disease or death, frequently due to renal damage [85, 86]. Fever, shivering, muscle tenderness, icterus, vomiting, anorexia, lethargy, abdominal pain, and changes in urination patterns such as polyuria, oliguria, or anuria are common clinical signs of canine leptospirosis [87]. Reports of clinical leptospirosis in cats are rare [88]. Clinical symptoms in wild animals vary with the host species, the pathogenicity of the *Leptospira* strain and serovar, as well as the age of the animal and physiological state [89]. Leptospirosis shows itself in a variety of wild animal species, with varying clinical manifestations. Infections that affect multiple mammalian species can cause systemic symptoms such as fever, as well as hepatic and renal involvement [90].

## 6. Diagnosis and detection in animals

A comprehensive approach to leptospirosis diagnosis that employs a variety of test modalities improves the accuracy of diagnosis. Increasing access to tests that meet regulatory and quality criteria for diagnosis is necessary to expand and encourage testing in animals in order to raise the global spotlight on leptospirosis. The proper diagnosis of animal leptospirosis is dependent on several critical aspects, including the collection of appropriate specimens and, most importantly, the stage of infection. Direct visualization of leptospires in blood, urine, and milk of infected animals has been performed using dark-field microscopy. However, false positives are observed due to the Brownian movement of fibrin proteins often misinterpreted as *Leptospira* [19], and this approach has low sensitivity and specificity [91]; therefore, it is not widely employed in the diagnosis of animal leptospirosis [16, 92]. A positive culture of biological specimens (blood, urine, tissue) provides definitive confirmation of infection; however, culturing leptospires is time-consuming and meticulous. The bacteria require a specific growth medium and may require months-long incubation. The culturing approach is highly specific but has limited sensitivity, which might be due to low bacteremia, antibiotic dosing prior to specimen collection, or technical issues [92].

Serologically, the microscopic agglutination test (MAT) is appropriate to diagnose acute leptospirosis, as infected animals have high IgM and IgG titers that are easily detectable, and its use is recommended by the WOAH [93]. The diagnostic performance of the MAT is strongly related to the number of serovars in the panel used for each region. This test, however, needs substantial knowledge to perform; thus, the results must be regarded with caution [94]. Other serological test kinds, such as IgM ELISA and other rapid diagnostic tests, such as indirect hemagglutination test, Latex agglutination assays, Lateral flow assays, and IgM Dipstick assays, have been developed as more accessible alternatives to the time-consuming MAT method [95]. Serologic assays for leptospirosis have several limitations. Although MAT is considered a "gold standard" reference serovar-specific test for leptospirosis diagnosis, it is challenging to perform correctly. It also requires a wide panel of live leptospiral culture as antigen, as well as professional expertise and qualified personnel for result interpretation. Similarly, ELISA and rapid diagnostic procedures that employ only one antigen to detect *Leptospira* spp. infection can fail to identify antibodies directed against additional antigens. Serology tests should ideally utilize an antigen (or antigens) that detects infection by all leptospiral species and strains precisely and sensitively [96]. Sero-diagnostic antigens should be updated on a regular basis when new strains appear to maximize their efficacy [95]. Despite its limitations as an early diagnostic test, serology has played an important role in leptospirosis monitoring. Sero-surveillance has shown the extent of leptospirosis exposure and identified high-risk animals and regions. MAT sero-surveillance, on the other hand, can give broad epidemiological information on circulating serovars [96, 97].

Molecular diagnostic methods for leptospiral DNA identification, such as PCR, have been widely used for animal leptospirosis diagnosis due to their excellent sensitivity and specificity [98]. The *16S rRNA* gene was the first genetic marker utilized for Leptospira identification and has been used as a target for various diagnostic PCR techniques for a long period [99]. Nowadays, PCR assays have focused on both universal genes present in bacteria such as *gryB*, *rrs* (*16S rRNA* gene), and *secY*, as well as surface proteins specific to *Leptospira* such as *lipL21*, *lipL32*, *lipL41,* and *ligB* [100]. *LipL32*-PCR, a surface protein gene, has been widely reported to be beneficial for initial screening for *Leptospira* DNA detection in animal clinical samples. Real-time PCR (qPCR) is now being used, which enables a faster and more sensitive diagnostic than classical PCR [101]. The limitation of PCR is that it can only quantify leptospires but not identify the specific strain, which is crucial in epidemiological surveillance [102].

## 7. Prevention and control measures in animals

Vaccination is an important step in preventing leptospirosis in animals. Standard vaccination protocols in dairy cattle herds, for example, normally begin with calves aged 4–6 weeks, followed by yearly or semi-annual boosters in closed or open herds, accordingly [103]. Vaccinations are advised for dogs around the world once a year [104]. Vaccinating animals against leptospirosis can aid in disease prevention. However, it is crucial to highlight that the vaccination does not give complete protection. This is due to the fact that there are several strains of leptospires, and the vaccination does not confer protection against all of them [105].

Biosecurity measures within barns and sites are critical in the context of bovines, pigs, and small ruminants such as goats and sheep. Rodent control measures are one of these strategies. Rodents serving as *Leptospira* reservoirs demand specific measures. Implementing rodent-proof structures and using rodenticides judiciously are critical in limiting the reservoir of the bacteria, hence significantly lowering the danger of intra-species transmission within animal populations. Furthermore, effective waste disposal techniques, avoiding contact with contaminated water sources, and enforcing strict quarantine measures for new introductions are critical components in limiting the introduction and subsequent spread of *Leptospira* within animal populations. Monitoring symptoms of infection and performing routine health checks that include serological testing are critical for early identification and effective control [106, 107].

Preventing leptospirosis in companion animals, particularly dogs with outdoor access, involves the execution of key interventions including selectively vaccinating against prevalent serovars like *Leptospira interrogans* serovar Canicola and *L. interrogans* serovar Icterohaemorrhagiae [108]. Other strategies include limiting exposure to contaminated water sources, decreasing rodent populations near homes, and assuring access to safe drinking water. Regular veterinarian examinations, including serological testing, are essential components of a complete preventive strategy. The complex dynamics introduced by wildlife populations, with their different species and environments, add complexity to leptospirosis prevention [109]. Strategies include monitoring for symptoms of infection, examining wildlife-human-animal interactions to reduce spillover occurrences, and contemplating selective vaccination, particularly in circumstances where certain animals serve as reservoirs [110]. The establishment of monitoring and surveillance networks assists in tracking leptospirosis prevalence and developing trends in animal populations. Regular testing and data collecting allows for informed decision-making, early detection of outbreaks, and focused response. In cases of confirmed leptospirosis cases, prompt and proper antibiotic therapy under veterinarian care is important. Antibiotics minimize the severity and duration of infection, hence reducing complications. Establishing treatment procedures ensures that affected animals are managed in a uniform and effective manner [111].

## 8. The one health approach

Recognizing the vital interdependence of human and animal health is critical in the context of leptospirosis prevention. The shared ecosystems, zoonotic potential, and cross-species transmission potential highlight the importance of coordinated efforts. This emphasizes the importance of a One Health approach, in which specialists in both human and animal health work together to address the varied challenges of disease prevention. Disease surveillance relies on the smooth collaboration of veterinarians and public health specialists. There is also the role of biologists, environmentalists, or ecologists, who monitor rodent populations, migrations of rodents depending on floods or droughts, and other parameters. This includes exchanging information on leptospirosis cases, coordinating control actions, and being committed to the One Health concept. This type of coordination improves early identification of epidemics, allowing for prompt intervention and accurate implementation of focused preventive measures. Effective leptospirosis management requires national and international collaboration. Sharing best practices and research findings, as well as coordinating efforts with surrounding regions, all contribute to a collective and coordinated approach to disease prevention, particularly in places with shared ecosystems and animal mobility.

## 9. Conclusion

Leptospirosis, as a zoonotic disease, endangers both animal and human health. As we understand more about the disease complexities, investing in ongoing research becomes a crucial step. This includes studying regional serovar prevalence, evaluating the effectiveness of current vaccinations, and looking for novel disease management techniques. The information gained from such studies not only improves the understanding of animal leptospirosis but also paves the road for more effective prevention and management strategies.

## Conflict of interest

The authors declare no conflict of interest regarding the content presented in this chapter.

*DOI: http://dx.doi.org/10.5772/intechopen.1008535*

## Author details

Durlav Prasad Bora[1*], Chandrani Goswami[1], Sophia Makdoh Gogoi[1], Arfan Ali[1], Nur Abdul Kader[1], Kalimuthusamy Natarajaseenivasan[2] and Nagendra Nath Barman[1]

1 Department of Veterinary Microbiology, College of Veterinary Science, Assam Agricultural University, Guwahati, Assam, India

2 Regional Medical Research Centre, Dibrugarh, Assam, India

*Address all correspondence to: drdpbora@gmail.com

IntechOpen

© 2025 The Author(s). Licensee IntechOpen. This chapter is distributed under the terms of the Creative Commons Attribution License (http://creativecommons.org/licenses/by/3.0), which permits unrestricted use, distribution, and reproduction in any medium, provided the original work is properly cited. (cc) BY

## References


[1] Levett PN. *Leptospira*. Manual of Clinical Microbiology. 11th ed. 2015. pp. 1028-1036. DOI: 10.1128/9781555817381.ch58

[2] Adler B, de la Peña MA. *Leptospira* and leptospirosis. Veterinary Microbiology. 2010;**140**:287-296. DOI: 10.1016/j.vetmic.2009.03.012

[3] Pal M, Roba Bulcha M, Mitiku BW. Leptospirosis and one health perspective. American Journal of Public Health Research. 2021;**9**:180-183. DOI: 10.12691/ajphr-9-4-9

[4] Rojas P, Monahan AM, Schuller S, Miller IS, Markey BK, Nally JE. Detection and quantification of leptospires in urine of dogs: A maintenance host for the zoonotic disease leptospirosis. European Journal of Clinical Microbiology and Infectious Diseases. 2010;**29**:1305-1309. DOI: 10.1007/s10096-010-0991-2

[5] Soo ZMP, Khan NA, Siddiqui R. Leptospirosis: Increasing importance in developing countries. Acta Tropica. 2020;**201**:105183. DOI: 10.1016/j.actatropica.2019.105183

[6] Guernier V, Lagadec E, Cordonin C, Le Minter G, Gomard Y, Pages F, et al. Human leptospirosis on Reunion Island, Indian Ocean: Are rodents the (only) ones to blame? PLoS Neglected Tropical Diseases. 2016;**10**:e0004733. DOI: 10.1371/journal.pntd.0004733

[7] Plank R, Dean D. Overview of the epidemiology, microbiology, and pathogenesis of *Leptospira* spp. in humans. Microbes and Infection. 2000;**2**:1265-1276. DOI: 10.1016/S1286-4579(00)01280-6

[8] Arean VM. The pathologic anatomy and pathogenesis of fatal human leptospirosis (Weil's disease). The American Journal of Pathology. 1962;**40**:393

[9] Kobayashi Y. Discovery of the causative organism of Weil's disease: Historical view. Journal of Infection and Chemotherapy. 2001;**7**:10-15. DOI: 10.1007/s101560170028

[10] Levett PN. Leptospirosis: A forgotten zoonosis? Clinical and Applied Immunology Reviews. 2004;**4**:435-448. DOI: 10.1016/j.cair.2004.08.001

[11] Saglam YS, Yener Z, Temur A, Yalcin E. Immunohistochemical detection of leptospiral antigens in cases of naturally occurring abortions in sheep. Small Ruminant Research. 2008;**74**:119-122. DOI: 10.1016/j.smallrumres.2007.04.006

[12] Bomfim MRQ, Koury MC. Evaluation of LSSP-PCR for identification of *Leptospira* spp. in urine samples of cattle with clinical suspicion of leptospirosis. Veterinary Microbiology. 2006;**118**:278-288. DOI: 10.1016/j.vetmic.2006.07.020

[13] Guitián FJ, García-Peña FJ, Oliveira J, Sanjuána ML, Yusa E. Serological study of the frequency of leptospiral infections among dairy cows in farms with suboptimal reproductive efficiency in Galicia, Spain. Veterinary Microbiology. 2001;**80**:275-284. DOI: 10.1016/S0378-1135(01)00306-6

[14] Tooloei M, Abdollapour G, Karimi H, Hasanpor A. Prevalence of serum antibodies against six *Leptospira* serovars in sheep in Tabriz, Northwestern Iran. Journal of Animal and Veterinary Advances. 2008;**3**:76. DOI: 10.6000/1927-520x.2014.03.03.2

[15] Clerke AM, Leuva AC, Joshi C, Trivedi SV. Clinical profile of leptospirosis in South Gujarat. Journal of Postgraduate Medicine. 2002;**48**:117-118

[16] Ellis WA. Animal Leptospirosis. 2015. DOI: 10.1007/978-3-662-45059-8_6

[17] World Organization for Animal Health (OIE). Terrestrial Manual Online Access. 2016

[18] Lilenbaum W, Varges R, Ristow P, Cortez A, Souza SO, Richtzenhain LJ, et al. Identification of *Leptospira* spp. carriers among seroreactive goats and sheep by polymerase chain reaction. Research in Veterinary Science. 2009;**87**:16-19. DOI: 10.1016/j.rvsc.2008.12.014

[19] Levett P. Leptospirosis. Clinical Microbiology Reviews. 2001;**14**:296-326

[20] Goldstein SF, Charon NW. Motility of the spirochete *leptospira*. Cell Motility and the Cytoskeleton. 1988;**9**:101-110. DOI: 10.1002/cm.970090202

[21] Wolff JW, Broom JC. The genus Leptospira Noguchi, 1917; problems of classification and a suggested system based on antigenic analysis. Documenta de medicina geographica et tropica. 1954;**6**:78-95

[22] Levett PN, Picardeau M. International committee on systematics of prokaryotes subcommittee on the taxonomy of Leptospiraceae. Minutes of the closed meeting, 28 November, Palmerston North, New Zealand. International Journal of Systematic and Evolutionary Microbiology. 2018;**68**:3362. DOI: 10.1099/ijsem.0.000815

[23] Adler B, Chappel R. International committee on systematics of prokaryotes subcommittee on the taxonomy of Leptospiraceae. International Journal of Systematic and Evolutionary Microbiology. 2011;**61**:698-699. DOI: 10.1099/ijs.0.031427-0

[24] Faine S, Adler B, Bolin C, Perolat P. *Leptospira* and Leptospirosis. 2nd ed. Melbourne, Australia: MediSci; 1999

[25] Bharti AR, Nally JE, Ricaldi JN, Matthias MA, Diaz MM, Lovett MA, et al. Leptospirosis: A zoonotic disease of global importance. Lancet Infectious Diseases. 2003;**3**:757-771. DOI: 10.1016/S1473-3099(03)00830-2

[26] Brenner DJ, Kaufmann AF, Sulzer KR, Steigerwalt AG, Rogers FC, Weyant RS. Further determination of DNA relatedness between serogroups and serovars in the family Leptospiraceae with a proposal for *Leptospira alexanderi sp*. nov. and four new *Leptospira* genomospecies. International Journal of Systematic Bacteriology. 1999;**Pt 2**:839-858. DOI: 10.1099/00207713-49-2-839

[27] Niyaz A, Manjulata SD, M de los Á V, P V, Robert S, Machangu WAE, Rudy AH. Multilocus sequence typing method for identification and genotypic classification of pathogenic *Leptospira* species. Annals of Clinical Microbiology and Antimicrobials. 2006;**5**:1-10. DOI: 10.1186/1476-0711-5-28

[28] Ramadass P, Jarvis BDW, Corner RJ, Penny D, Marshall RB. Genetic characterization of pathogenic *Leptospira* species by DNA hybridization. International Journal of Systematic Bacteriology. 1992;**42**:215-219. DOI: 10.1099/00207713-42-2-215

[29] Yasuda PH, Steigerwalt AG, Sulzer KR, Kaufmann AF, Rogers F, Brenner DJ. Deoxyribonucleic acid relatedness between serogroups and serovars in the family leptospiraceae with proposals for seven new *Leptospira* species. International Journal of

Systematic Bacteriology. 1987;**37**:407-415. DOI: 10.1099/00207713-37-4-407

[30] Levett PN. Systematics of leptospiraceae. Current Topics in Microbiology and Immunology. 2015;**387**:11-20. DOI: 10.1007/978-3-662-45059-8_2

[31] Goarant C. Leptospirosis: Risk factors and management challenges in developing countries. Research and Reports in Tropical Medicine. 2016;7:49-62. DOI: 10.2147/rrtm.s102543

[32] Cosson JF, Picardeau M, Mielcarek M, Tatard C, Chaval Y, Suputtamongkol Y, et al. Epidemiology of Leptospira transmitted by rodents in Southeast Asia. PLoS Neglected Tropical Diseases. 2014;**8**:e2902. DOI: 10.1371/journal.pntd.0002902

[33] Costa F, Wunder EA, de Oliveira D, Bisht V, Rodrigues G, Reis MG, et al. Patterns in Leptospira shedding in Norway rats (*Rattus norvegicus*) from Brazilian slum communities at high risk of disease transmission. PLoS Neglected Tropical Diseases. 2015;**9**:e0003819. DOI: 10.1371/journal.pntd.0003819

[34] Dietrich M, Wilkinson DA, Benlali A, Lagadec E, Ramasindrazana B, Dellagi K, et al. Leptospira and paramyxovirus infection dynamics in a bat maternity enlightens pathogen maintenance in wildlife. Environmental Microbiology. 2015;**17**:4280-4289. DOI: 10.1111/1462-2920.12766

[35] Lagadec E, Gomard Y, Le Minter G, Cordonin C, Cardinale E, Ramasindrazana B, et al. Identification of Tenrec ecaudatus, a wild mammal introduced to Mayotte Island, as a reservoir of the newly identified human pathogenic *Leptospira mayottensis*. PLoS Neglected Tropical Diseases. 2016;**10**:e0004933. DOI: 10.1371/journal.pntd.0004933

[36] Dietrich M, Wilkinson DA, Soarimalala V, Goodman SM, Dellagi K, Tortosa P. Diversification of an emerging pathogen in a biodiversity hotspot: *Leptospira* in endemic small mammals of Madagascar. Molecular Ecology. 2014;**23**:2783-2796. DOI: 10.1111/mec.12777

[37] Barragan V, Chiriboga J, Miller E, Olivas S, Birdsell D, Hepp C, et al. High Leptospira diversity in animals and humans complicates the search for common reservoirs of human disease in rural Ecuador. PLoS Neglected Tropical Diseases. 2016;**10**:e0004990. DOI: 10.1371/journal.pntd.0004990

[38] Hamir AN, Hanlon CA, Niezgoda M, Rupprecht CE. The prevalence of interstitial nephritis and leptospirosis in 283 raccoons (Procyon lotor) from 5 different sites in the United States. Canadian Veterinary Journal. 2001;**42**:869-871

[39] Bengis RG, Leighton FA, Fischer JR, Artois M, Mörner T, Tate CM. The role of wildlife in emerging and re-emerging zoonoses. OIE Revue Scientifique et Technique. 2004;**23**:497-512. DOI: 10.20506/rst.23.2.1498

[40] Crecelius EM, Burnett MW. Leptospirosis. Journal of Special Operations Medicine: A Peer Reviewed Journal for SOF Medical Professionals. 2020;**20**:121-122. DOI: 10.15581/021.50.7603

[41] Lehmann JS, Matthias MA, Vinetz JM, Fouts DE. Leptospiral pathogenomics. Pathogens. 2014;**3**:280-308. DOI: 10.3390/pathogens3020280

[42] Constable P, Hinchcliff K, Done S, Grünberg W. Veterinary medicine. Text book of the diseases of cattle, horses, sheep, pigs and goats edition 11. Elsevier Health Sciences. 2016;**1**:1115

[43] Cordonin C, Turpin M, Bringart M, Bascands JL, Flores O, Dellagi K, et al. Pathogenic *Leptospira* and their animal reservoirs: Testing host specificity through experimental infection. Scientific Reports. 2020;**10**:723. DOI: 10.1038/s41598-020-64172-4

[44] Ko AI, Goarant C, Picardeau M. *Leptospira*: The dawn of the molecular genetics era for an emerging zoonotic pathogen. Nature Reviews Microbiology. 2009;7:736-747. DOI: 10.1038/nrmicro2208

[45] Pépin M, Dupinay T, Zilber AL, McElhinney LM. Of rats and pathogens: Pathogens transmitted by urban rats with an emphasis on hantaviruses. CAB Reviews: Perspectives in Agriculture, Veterinary Science, Nutrition and Natural Resources. 2016:1-13. DOI: 10.1079/PAVSNNR201611019. Available from: https://www.cabidigitallibrary.org/doi/10.1079/PAVSNNR201611019#indexing-terms

[46] Dhanze H, Kumar S, MBG. Epidemiology of leptospirosis: An Indian perspective. Journal of Foodborne and Zoonotic Diseases. 2013;**1**:6-13

[47] Yadeta W, Michael BG, Abdela N, Box PO. Leptospirosis in animal and its public health implications: A review. World Apllied Sciences Journal. 2016;**34**:845-853. DOI: 10.5829/idosi.wasj.2016.34.6.103113

[48] Zucca P, Scagliarini A, Ramma Y, Khan AS. Zoonoses—Diseases naturally transmitted from animals to humans. Frontiers for Young Minds. 2022;**10**:1-8. DOI: 10.3389/frym.2022.833893

[49] Marrana M. Epidemiology of disease through the interactions between humans, domestic animals, and wildlife. In: One Health: Integrated Approach to 21st Century Challenges to Health. 2022. pp. 73-111. DOI: 10.1016/B978-0-12-822794-7.00001-0

[50] Esposito MM, Turku S, Lehrfield L, Shoman A. The impact of human activities on zoonotic infection transmissions. Animals. 2023;**13**:1646. DOI: 10.3390/ani13101646

[51] Munoz-Zanzi C, Groene E, Morawski BM, Bonner K, Costa F, Bertherat E, et al. A systematic literature review of leptospirosis outbreaks worldwide, 1970-2012. Revista Panamericana de Salud Publica/Pan American Journal of Public Health. 2020;**44**:e78. DOI: 10.26633/RPSP.2020.78

[52] Casanovas-Massana A, Pedra GG, Wunder EA, Diggle PJ, Begon M, Ko AI. Quantification of *Leptospira* interrogans survival in soil and water microcosms. Applied and Environmental Microbiology. 2018;**84**:e00507. DOI: 10.1128/AEM.00507-18

[53] Thibeaux R, Iraola G, Ferrés I, Bierque E, Girault D, Soupé-Gilbert ME, et al. Deciphering the unexplored *Leptospira* diversity from soils uncovers genomic evolution to virulence. Microbial Genomics. 2018;**4**:e000144. DOI: 10.1099/mgen.0.000144

[54] Ismail S, Wahab NZA, Badya N, Rahman NI, Yeo CC, Latif AZ, et al. A study on the presence of pathogenic *Leptospira* spp. in environmental water samples obtained from selected recreational areas in Terengganu, Malaysia. Research Journal of Pharmacy and Technology. 2014;7:1153-1157

[55] Diesch SL, McCulloch WF. Isolation of pathogenic leptospires from waters used for recreation. Public Health Reports. 1966;**81**:299. DOI: 10.2307/4592702

[56] Wilson R, Fujioka R. Development of a method to selectively isolate pathogenic *Leptospira* from environmental samples. Water Science

and Technology. 1995;**31**:275-282. DOI: 10.1016/0273-1223(95)00280-Z

[57] Lall C, Kumar KV, Raj RV, Vedhagiri K, Sunish IP, Vijayachari P. Correlation between physicochemical properties of soil and presence of *Leptospira*. EcoHealth. 2018;**15**:670-675. DOI: 10.1007/s10393-018-1346-1

[58] Wójcik-Fatla A, Zajac V, Wasinski B, Sroka J, Cisak E, Sawczyn A, et al. Occurrence of *Leptospira* DNA in water and soil samples collected in eastern Poland. Annals of Agricultural and Environmental Medicine. 2014;**21**:730-732. DOI: 10.5604/12321966.1129924

[59] Hacker KP, Sacramento GA, Cruz JS, de Oliveira D, Nery N Jr, Lindow JC, et al. Influence of rainfall on *Leptospira* infection and disease in a tropical urban setting, Brazil. Emerging Infectious Diseases. 2020;**26**:311. DOI: 10.3201/eid2602.190102

[60] Wynwood SJ, Graham GC, Weier SL, Collet TA, McKay DB, Craig SB. Leptospirosis from water sources. Pathogens and Global Health. 2014;**108**:334-338. DOI: 10.1179/2047773214Y.0000000156

[61] Chadsuthi S, Chalvet-Monfray K, Wiratsudakul A, Modchang C. The effects of flooding and weather conditions on leptospirosis transmission in Thailand. Scientific Reports. 2021;**11**:1486. DOI: 10.1038/s41598-020-79546-x

[62] Builes-Jaramillo A, Arias-Monsalve CS. Impact of El Niño-southern oscillation on human leptospirosis in Colombia at different spatial scales. Journal of Infection in Developing Countries. 2019;**13**:1108-1116. DOI: 10.3855/jidc.11702

[63] Weinberger D, Baroux N, Grangeon JP, Ko AI, Goarant C. El Niño southern oscillation and leptospirosis outbreaks in New Caledonia. PLoS Neglected Tropical Diseases. 2014;**8**:e2798. DOI: 10.1371/journal.pntd.0002798

[64] Pawar S, Kore M, Athalye A, Thombre P. Seasonality of leptospirosis and its association with rainfall and humidity in Ratnagiri, Maharashtra. International Journal of Health and Allied Sciences. 2018;**7**:37-40. DOI: 10.4103/ijhas.ijhas_35_16

[65] Cunha M, Costa F, Ribeiro GS, Carvalho MS, Reis RB, Nery N Jr, et al. Rainfall and other meteorological factors as drivers of urban transmission of leptospirosis. PLoS Neglected Tropical Diseases. 2022;**16**:e0007507. DOI: 10.1371/journal.pntd.0007507

[66] Mohd Radi MF, Hashim JH, Jaafar MH, Hod R, Ahmad N, Nawi AM, et al. Leptospirosis outbreak after the 2014 major flooding event in Kelantan, Malaysia: A spatial-temporal analysis. American Journal of Tropical Medicine and Hygiene. 2018;**98**:1281-1295. DOI: 10.4269/ajtmh.16-0922

[67] Rahman T, Candolin U. Linking animal behavior to ecosystem change in disturbed environments. Frontiers in Ecology and Evolution. 2022;**10**:893453. DOI: 10.3389/fevo.2022.893453

[68] Browne ES, Callefe JLR, Jesus DE, ERS, Zeppelini CG, Cremonese C, Costa F. A systematic review of the geographic distribution of pathogenic *Leptospira* serovars in the Americas, 1930-2017. Anais da Academia Brasileira de Ciencias. 2022;**94**:e20201026. DOI: 10.1590/0001-3765202220201026

[69] Dewes C, Silva JPM, Fortes TP, Marmitt IV, Vasconcellos FA, Felix SR, et al. Prevalence of Anti-*Leptospira* Antibodies in Cattle Confined for

Exportation. Research, Society and Development. 2020;**9**:e3329119929. DOI: 10.33448/rsd-v9i11.9929

[70] Otaka DY, Martins G, Hamond C, Penna B, Medeiros MA, Lilenbaum W. Serology and PCR for bovine leptospirosis: Herd and individual approaches. Veterinary Record. 2012;**170**:338. DOI: 10.1136/vr.100490

[71] Mgode GF, Machang'u RS, Mhamphi GG, Katakweba A, Mulungu LS, Durnez L, et al. *Leptospira* serovars for diagnosis of leptospirosis in humans and animals in Africa: Common *Leptospira* isolates and reservoir hosts. PLoS Neglected Tropical Diseases. 2015;**9**:e0004251. DOI: 10.1371/journal.pntd.0004251

[72] Sohm C, Steiner J, Jöbstl J, Wittek T, Firth C, Steinparzer R, et al. A systematic review on leptospirosis in cattle: A European perspective. One Health. 2023;**17**:100608. DOI: 10.1016/j.onehlt.2023.100608

[73] Mgode GF, Mhamphi GG, Massawe AW, Machang'u RS. *Leptospira* seropositivity in humans, livestock and wild animals in a semi-arid area of Tanzania. Pathogens. 2021;**10**:696. DOI: 10.3390/pathogens10060696

[74] Rahman MS, Bejo SK, Zakaria Z, Hassan L, Roslan MA. Seroprevalence and distribution of leptospiral serovars in livestock (cattle, goats, and sheep) in flood-prone Kelantan, Malaysia. Journal of Veterinary Research. 2021;**65**:53-58. DOI: 10.2478/jvetres-2021-0003

[75] Roquelo C, Kodjo A, Lou MJ, Davoust B. Serological and molecular survey of *Leptospira* spp. infections in wild boars and red foxes from southeastern France. Veterinary World. 2021;**14**:825. DOI: 10.14202/VETWORLD.2021.825-828

[76] Saranya P, Goswami C, Sumathi K, Balasundareshwaran AH, Bothammal P, Dutta LJ, et al. Prevalence of leptospirosis among animal herds of north eastern provinces of India. Comparative Immunology, Microbiology and Infectious Diseases. 2021;**79**:101698. DOI: 10.1016/j.cimid.2021.101698

[77] Ikpe AB, Ahmed K, Sarma DK, Bora DP, Natarajaseenivasan K, Goswami C, et al. Seroprevalence of leptospirosis in dairy cows with reproductive disorders. Journal of Animal Research. 2021;**11**:801-805. DOI: 10.30954/2277-940x.05.2021.5

[78] Victoriano AFB, Smythe LD, Gloriani-Barzaga N, Cavinta LL, Kasai T, Limpakarnjanarat K, et al. Leptospirosis in the Asia Pacific region. BMC Infectious Diseases. 2009;**9**:1-9. DOI: 10.1186/1471-2334-9-147

[79] Senthil NR, Palanivel KM, Rishikesavan R. Seroprevalence of Leptospiral antibodies in canine population in and around Namakkal. Journal of Veterinary Medicine. 2013;**2013**:1-4. DOI: 10.1155/2013/971810

[80] Chadsuthi S, Bicout DJ, Wiratsudakul A, Suwancharoen D, Petkanchanapong W, Modchang C, et al. Investigation on predominant *Leptospira* serovars and its distribution in humans and livestock in Thailand, 2010-2015. PLoS Neglected Tropical Diseases. 2017;**11**:e0005228. DOI: 10.1371/journal.pntd.0005228

[81] Laras K, Van CB, Bounlu K, Nguyen TK, Olson JG, Thongchanh S, et al. The importance of leptospirosis in Southeast Asia. American Journal of Tropical Medicine and Hygiene. 2002;**67**:278-286. DOI: 10.4269/ajtmh.2002.67.278

[82] Guedes IB, de Souza GO, de Oliveira LAR, Castro JF, de Souza

Filho AF, Maia AL, et al. Prevalence of *Leptospira* serogroups in buffaloes from the Brazilian Amazon. Veterinary Medicine and Science. 2020;**6**:433-440. DOI: 10.1002/vms3.271

[83] Ibrahim NA, Alrashdi BM, Elnaker YF, Elmahallawy EK, Alblihed MA, Daib MS, et al. Serological investigation and epidemiological analysis of bovine leptospirosis in Egypt. Tropical Medicine and Infectious Disease. 2022;**7**:208. DOI: 10.3390/tropicalmed7090208

[84] Orjuela AG, Parra-Arango JL, Sarmiento-Rubiano LA. Bovine leptospirosis: Effects on reproduction and an approach to research in Colombia. Tropical Animal Health and Production. 2022;**54**:251. DOI: 10.1007/s11250-022-03235-2

[85] Murphy K. Dealing with leptospirosis in dogs. Veterinary Record. 2018;**183**:384. DOI: 10.1136/vr.k4093

[86] Furlanello T, Reale I. Leptospirosis and immune-mediated hemolytic anemia: A lethal association. Veterinary Research Forum. 2019;**10**:261. DOI: 10.30466/vrf.2019.99876.2385

[87] Andre-Fontaine G. Canine leptospirosis—Do we have a problem? Veterinary Microbiology. 2006;**117**:19-24. DOI: 10.1016/j.vetmic.2006.04.005

[88] Mazzotta E, De Zan G, Cocchi M, Boniotti MB, Bertasio C, Furlanello T, et al. Feline susceptibility to leptospirosis and presence of immunosuppressive co-morbidities: First European report of L. interrogans serogroup Australis sequence type 24 in a cat and survey of *Leptospira* exposure in outdoor cats. Tropical Medicine and Infectious Disease. 2023;**8**:54. DOI: 10.3390/tropicalmed8010054

[89] Cilia G, Bertelloni F, Albini S, Fratini F. Insight into the epidemiology of leptospirosis: A review of *Leptospira* isolations from "unconventional" hosts. Animals. 2021;**11**:1-16. DOI: 10.3390/ani11010191

[90] Hu W, Lin X, Yan J. *Leptospira* and leptospirosis in China. Current Opinion in Infectious Diseases. 2014;**27**:432-436. DOI: 10.1097/QCO.0000000000000097

[91] Jaiswal NK, Singh BK, Chandrasekaran S, Kumar R. Evaluation of dark field microscopy for the early diagnosis of leptospirosis-A prospective study. Journal of Medical Science and Clinical Research. 2018;**6**:798-801. DOI: 10.18535/jmscr/v6i8.133

[92] Reagan KL, Sykes JE. Diagnosis of canine leptospirosis. Veterinary Clinics of North America—Small Animal Practice. 2019;**49**:719-731. DOI: 10.1016/j.cvsm.2019.02.008

[93] OIE. Manual of Diagnostic Tests and Vaccines for Terrestrial Animals. 7th ed. Paris: World Organization for Animal Health; 2012

[94] Loureiro AP, Lilenbaum W. Genital bovine leptospirosis: A new look for an old disease. Theriogenology. 2020;**141**:41-47. DOI: 10.1016/j.theriogenology.2019.09.011

[95] Picardeau M, Bertherat E, Jancloes M, Skouloudis AN, Durski K, Hartskeerl RA. Rapid tests for diagnosis of leptospirosis: Current tools and emerging technologies. Diagnostic Microbiology and Infectious Disease. 2014;**78**:1-8. DOI: 10.1016/j.diagmicrobio.2013.09.012

[96] Sykes JE, Reagan KL, Nally JE, Galloway RL, Haake DA. Role of diagnostics in epidemiology, management, surveillance, and control

of leptospirosis. Pathogens. 2022;**11**:395. DOI: 10.3390/pathogens11040395

[97] Subharat S, Wilson PR, Heuer C, Collins-Emerson JM, Smythe LD, Dohnt MF, et al. Serosurvey of leptospirosis and investigation of a possible novel serovar Arborea in farmed deer in New Zealand. New Zealand Veterinary Journal. 2011;**59**:139-142. DOI: 10.1080/00480169.2011.561780

[98] Hamond C, Martins G, Bremont S, Medeiros MA, Bourhy P, Lilenbaum W. Predominance of *Leptospira* interrogans serovar Bratislava DNA in vaginal fluid of mares suggests sexual transmission of leptospirosis. Animal Reproduction Science. 2014;**151**:275-279. DOI: 10.1016/j.anireprosci.2014.10.019

[99] Smythe LD, Smith IL, Smith GA, Dohnt MF, Symonds ML, Barnett LJ, et al. A quantitative PCR (TaqMan) assay for pathogenic *Leptospira* spp. BMC Infectious Diseases. 2002;**2**:1-7. DOI: 10.1186/1471-2334-2-13

[100] Guernier V, Allan KJ, Goarant C. Advances and challenges in barcoding pathogenic and environmental *Leptospira*. Parasitology. 2018;**145**:595-607. DOI: 10.1017/S0031182017001147

[101] Picardeau M. Diagnosis and epidemiology of leptospirosis. Medecine et Maladies Infectieuses. 2013;**43**:1-9. DOI: 10.1016/j.medmal.2012.11.005

[102] Haake DA, Levett PN. Leptospirosis in humans. Current Topics in Microbiology and Immunology. 2015;**387**:65-97. DOI: 10.1007/978-3-662-45059-8_5

[103] Mughini-Gras L, Bonfanti L, Natale A, Comin A, Ferronato A, La Greca E, et al. Application of an integrated outbreak management plan for the control of leptospirosis in dairy cattle herds. Epidemiology and Infection. 2014;**142**:1172-1181. DOI: 10.1017/S0950268813001817

[104] Day MJ, Horzinek MC, Schultz RD, Squires RA. WSAVA guidelines for the vaccination of dogs and cats. Journal of Small Animal Practice. 2016;**57**:E1-E45. DOI: 10.1111/jsap.2_12431

[105] Adler B. Vaccines against leptospirosis. Current Topics in Microbiology and Immunology. 2015;**387**:251-272. DOI: 10.1007/978-3-662-45059-8_10

[106] Minter A, Costa F, Khalil H, Childs J, Diggle P, Ko AI, et al. Optimal control of rat-borne leptospirosis in an urban environment. Frontiers in Ecology and Evolution. 2019;7:209. DOI: 10.3389/fevo.2019.00209

[107] Sahlström L, Virtanen T, Kyyrö J, Lyytikäinen T. Biosecurity on Finnish cattle, pig and sheep farms—Results from a questionnaire. Preventive Veterinary Medicine. 2014;**117**:59-67. DOI: 10.1016/j.prevetmed.2014.07.004

[108] Barazzone GC, Teixeira AF, Azevedo BOP, Damiano DK, Oliveira MP, Nascimento AL, et al. Revisiting the development of vaccines against pathogenic *Leptospira*: Innovative approaches, present challenges, and future perspectives. Frontiers in Immunology. 2022;**12**:760291. DOI: 10.3389/fimmu.2021.760291

[109] Montalvo CM. Leptospirosis at human-animal-environment interfaces in Latin-America: Drivers, prevention, and control measures. Biotecnia. 2021;**23**:89-100. DOI: 10.18633/biotecnia.v23i3.1442

[110] Vora NM, Hannah L, Walzer C, Vale MM, Lieberman S, Emerson A, et al. Interventions to reduce risk for pathogen spillover and early disease

spread to prevent outbreaks, epidemics, and pandemics. Emerging Infectious Diseases. 2023;**29**:E1-E9. DOI: 10.3201/eid2903.221079

[111] Guzmán Pérez M, Blanch Sancho JJ, Segura Luque JC, Mateos Rodriguez F, Martínez Alfaro E, Del Pozo SG, et al. Current evidence on the antimicrobial treatment and chemoprophylaxis of human leptospirosis: A meta-analysis. Pathogens. 2021;**10**:1125. DOI: 10.3390/pathogens10091125

Chapter 6

# Leptospirosis and the Environment

*Patrick Banza*

## Abstract

This study on leptospirosis and the environment aims to highlight the role of bacteria in the environment and the impact of the environment on *Leptosira* infections and human activities. The environment is a highly complex ecosystem in which many microorganisms develop and create forms of resistance. The environment is crucial to human activity and survival. If this environment is not well protected, it will be responsible for the modification of many non-pathogenic microorganisms into organisms that are pathogenic to mankind. If a microorganism finds a favorable environment or changes its natural habitat, it becomes a pathogen and through the phenomenon of bacterial conjugation, we can have a new bacteria that can escape the control of the environment.

**Keywords:** environment, bacteria, spirochetes, prevention, leptospirosis

## 1. Introduction

Microorganisms are not alone in nature. Each microorganism in an ecosystem interacts with everything around it and with other organisms [1]. These interactions can lead to significant physical and chemical changes in the environment, both beneficial and detrimental to other organisms. Whether harmful, beneficial, or without effect, the collective activities of microorganisms largely control the functioning of the biosphere [2].

In this chapter, the essential microbial activities will be analyzed.

First, spirochetes in general will be looked at, as the genus Leptospira, the various natural habitats of microorganisms, and then some specific environments will be explored: Water and soil are the two most important potential sources of the disease, and their purification and treatment is the only way to ensure public health.

Most frequently, the determination of water quality is done, which often depends on standardized microbiological techniques. For soil these analyses are less frequent, except among farmers and livestock breeders [3].

## 2. The pathogen

### 2.1 General information and descriptions of *spirochetes*

Main genera of *Sprochetes* are *Sprirochaeta, Treponema, Cristispira, Leptospira* and *Borrelia* [4].

Spirochaetes are gram-negative, motile bacteria with a slender, flexible helix. These prokaryotes, with their distinctive morphology, form a major phylogenetic division within the bacterial family. Spirochetes are widespread in aquatic environments and in animals. Some are the cause of diseases such as syphilis, in humans, and others such as leptospirosis [4].

Spirochetes are characterized by the protoplasmic cylindrical body, containing cytoplasm and nucleoid, which is delimited by a plasma membrane and a wall and corresponds to the body of the cell. The locomotor apparatus is internal. It is made up of one or more flagella, one end of each of which attaches to a pole of the protoplasmic cylinder, while the other remains located in the cell's periplasm, hence the name endo-flagellum. An outer membrane surrounds the endo-flagellae and the protoplasmic cylinder [5].

Spirochetes have a particular mobility. Each endo-flagellum is anchored at one end of the cell and runs two-thirds of the length of the protoplasmic cylinder. Endo-flagella rotates in the same way as flagella in other bacteria. However, as the protoplasmic cylinder is rigid, while the outer mantle is flexible, when the endo-flagellae at either end rotates in the same direction, the protoplasmic cylinder rotates in the opposite direction, twisting the cell. This gives spirochetes mobility by flexion, or creep, due to the torque exerted at each end of the protoplasmic cylinder by the rotating endo-flagellae [5].

Spirochaetes are divided into eight genera based on their habitat, pathogenicity, morphological and physiological characteristics. These are *Cristipira, Spirochaeta, Treponema, Borrelia, Leptosira, leptonema, Brachyspira* and *Brevinema*. Only *Leptospira* genus will be discussed.

## 3. Leptospira

### 3.1 Epidemiology of the pathogen

Although the global burden of the disease is unknown, ≥500,000 cases of leptospirosis are estimated to occur worldwide each year. The incidence in some areas may be as high as 975 cases per 100,000 population. During the past decade, the occurrence of outbreaks has highlighted the strong links between leptospirosis and extreme weather events in Guyana, India, Kenya, the Lao People's Democratic Republic, New Caledonia, Nicaragua, the Philippines and Thailand [6]. There is little research on the economic impact of leptospirosis, and information is lacking about the societal costs of the disease, including the costs of health care, lost productivity caused by sequelae and death of livestock. Two groups are considered to be most at risk: people living in urban slums and subsistence farmers [6]. The relative proportions of these groups in the population vary from country to country. There is often a constant force of infection owing to the existence of animal reservoirs, including rodents, livestock and dogs. In some areas, such as parts of Asia, outbreaks most often occur after flooding, as happened in the Philippines in November 2009 [6].

*Leptospira interrogans* is a Gram-negative, strictly aerobic bacterium responsible for causing the infectious disease leptospirosis, also known as Weil's disease, canicola fever or mud fever.

*Leptospira* cell is slender, shaped like a fine helix and generally bent at each end into a semicircular hook. This genus includes both free-living and parasitic species. The two major species are *L. interrogans* (parasitic and pathogen) and *Leptospira biflexa* (free non-pathogen).

*Leptospira* have a typical double membrane, with a lipo-polysaccharide composition of the outer membrane similar to that of gram-negative bacteria, but with a less marked endo-toxic activity (e.g., Leptospira lipopolysaccharides (LPS) are about 12 times less lethal to mice than *E. coli* LPS). A certain resemblance to Gram-positive bacteria also lies in the fact that the peptidoglycan layer is intimately associated with the cytoplasmic membrane [7].

These bacteria are therefore sensitive to most antibiotics but do not take the GRAM stain. These bacteria remain visible only under dark-field microscopy after silver staining or artificial thickening by immunoperoxidase or immunofluorescence [4].

The outer membrane plays an important role in *Leptospira*-induced immunity. In fact, preparations of leptospiral LPS are sufficient to induce immunity, but this is often serovar-dependent. Differences in the structure and composition of these LPS are thought to be responsible for the wide diversity of serovars. More recent studies have even shown that combinations of recombinant outer membrane proteins or 28 outer membrane preparations without LPS are also sufficient to induce immunity in laboratory animals, even conferring heterologous-type protection [2, 3, 5, 7–21].

The *Leptospira* genera comprise strict aerobic spirochaetes using long-chain fatty acids (e.g., oleic acid) as electron donors and carbon sources. With a few exceptions, these are the only substrates used by leptospires for growth [2, 3, 5, 8–17].

Leptospires are strict aerobes, catalase-positive and/or oxidase-positive and chemoorganotrophs. The optimum growth temperature is between 28 and 32°C. The environment in which leptospires grow must contain certain specific elements:

- Long-chain fatty acids (at least 14 carbon atoms), the only source of carbon, which is then metabolized by ß-oxidation.
- Ammonium salts, a source of nitrogen in inorganic form.
- Vitamins, notably B1 and B12, which are growth factors, as well as $Ca^{2+}$, $Mg^{2+}$ and $Fe^{2+}$ ions [2–5, 7–22].

*L. interrogans* strains are pathogens for humans and animals. The natural hosts are rodents, although dogs and pigs are carriers of certain strains.

Species likely to become ill and which also eliminate the bacterium *via* their urine are various domestic animals (dogs, horses, cattle and pigs) and wild animals (foxes, deer, wild boar, etc.).

Water and soil contaminated by the urine of infected animals are mostly sources of infection. *Leptospira* can live for a long time in soil and water (mostly sewers and mines).

The bacterial genome is made up of nitrogenous bases. These bases can undergo several reactions in nature, which can modify the virulence of the leptospire.

Bacteria and certain archaea possess an enzyme called methyltransferase, which is responsible for the methylation reaction; a variation in the gene encoding this enzyme would have serious consequences for the bacteria [1, 23].

In humans, leptospirosis is the most common syndrome caused by *Leptospira interrogans*.

Leptospires enter the body *via* mucous membranes or skin lesions. After a transitory stage of multiplication in various parts of the body, bacteria localizes in the kidney and liver, causing nephritis and jaundice [8].

**Figure 1.**
*Example of the sign for rat elimination obligation.*

Leptospira in the kidney causes kidney failure, which can lead to death. The microorganism is eliminated from the body through urine, and infection of another individual occurs through contact with contaminated urine.

Therapy with penicillin, streptomycin or tetracycline is available, but a long course of treatment is needed to ensure complete elimination of leptospires from the kidney. Domestic animals such as dogs are vaccinated against leptospirosis with the combined distemper-leptospirosis-hepatitis vaccine. In humans, prevention involves eliminating the disease from animals and eliminating rodents as potential sources of the infection (**Figure 1**) [5].

### 3.2 Genomic classification

A genomospecies is defined as "strains with 70% or more DNA-DNA homology and with a ΔTm of 5°C or less." The proximity of strains revealed by the MAT, which enables serogroups to be defined, does not always reflect genomic proximity: serovars belonging to the same serogroup are dispersed within several species, and a single species may comprise numerous serogroups. Brenner et al. [11] showed that out of 303 leptospires strains studied, representing 268 serovars grouped into 30 serogroups, only strains of the serogroup. In the genomic classification, the subgroup of saprophytic species, of which there are six, forms the oldest branch of the genus *Leptospira*. Another subgroup includes the nine pathogenic species, while another branch of evolution, of more recent discovery, includes the so-called intermediate species, for which pathogenicity has not been clearly demonstrated [16]. Although *L. kmeyti* was isolated from soil in Malaysia [20], phylogenetic analysis of 16S rRNA shows that this strain belongs to the cluster of pathogenic species. DNA variability between species is between 0.1 and 1.3% for pathogenic species, between 0.3 and 2.9% for intermediate species, and between 0.2 and 0.5% for pathogenic species. It suggests that the geographical distribution of the species is limited. For example, the six strains identified of *L. alexanderi* were all isolated in China. Sixty-two of the 65 strains of *L. santarosai* come from North and South America, as do the 20 strains of *L. noguchii*. Almost half of the 91 strains of *L. interrogans* were isolated in Oceania (representing 70% of the strains isolated in this region). Finally, the 37 African strains (characterized by DNA–DNA hybridization) in the study by Feresu et al. [13] all belong to either *L. kirschneri* or *L. santarosai*. Genomic and antigenic classification are not concordant, and several studies have shown that serogroups cannot be superimposed on groups formed by genetic classification [4], suggesting that the genes that determine serotypes could be transferred laterally between the different species. Thus, a classification system based on genetic similarities is used in conjunction with antigenic classification. Recently, the sequencing of the genome of strains belonging to several *Leptospira* species has enabled the introduction of various *Leptospira* genotyping approaches, including MLVA or MLST [9].

### 3.3 Cycles of sylvatic transmission (rodents)

Rodents are closely associated with human habitation and agriculture, mainly because they provide food and shelter for animals. Transmission of *Leptospira spp*. in rodent populations is governed by a complex interplay of rodent population dynamics, depending on food availability, climatic conditions and endemic status. In certain circumstances, the occurrence of leptospirosis is highly seasonal, with peaks in transmission occurring at the same time as seasonal increases in rainfall in tropical and subtropical areas (i.e., the rainy season) [14]. This type of human disease has been correlated with an increase in the size of the rodent population and an increase in the frequency of transport (excretion) of leptospires. This association can be explained by the increase in the survival rate of leptospires in natural environments, which favored an increase in the infection rate of rodents at the same time as the increase in the rodent population (probably attributable to an increase in available food and housing) created a larger pool of naïve individuals [2, 3, 5, 7, 14–19]. The epidemiology and in particular the seasonality of leptospirosis differs between the various agro-ecological zones in Fiji. There is a more apparent seasonal pattern in the north-west of the main island, which is associated with heavy seasonal rainfall and frequent flooding. The pattern of disease occurrence in the southeastern part of the main island is less seasonal, which may be due to a more even distribution of rainfall or to a different type of reservoir host.

## 4. The environment

One of those diseases is leptospirosis, which can be listed as an environmental disease. *Leptospira* survives provided that the environment is favorable for the adaptation of its metabolism (strict aerobic), thanks to the availability of growth factors.

Fresh or humid water contains a concentration of dissolved oxygen, which varies between 4.5 to 5.6 mg/l for an average temperature 30°C higher, enough to allow its survival [24].

The soil as a microbial environment depends on climate and biological processes, which determine the recycling of organic and inorganic compounds.

Microorganisms inhabit all habitats capable of supporting life. Microbial habitats are extremely diverse and often change very rapidly. In addition to the common habitats of soil and water, microorganisms live on the surface of more complex organisms and, in many cases, live inside them. Microorganisms frequently reach very large numbers in these habitats and can contribute significantly to the nutritional needs of plants or animals [4].

The presence of rodents in livestock buildings entails serious risks for farmers due to the number of pathogens that they carry. Rodents such as rats and mice, by carrying pathogens, represent a danger of different diseases that is often underestimated by the breeders.

One of those diseases is leptospirosis, which can be listed as an environmental disease [2, 3, 5, 7, 13–21].

Leptospira interrogans uses certain long-chain fatty acids such as oleic acid, as a carbon source [4].

Oleic acid is an 18-carbon fatty acid with a double bond that is monounsaturated in cis form. Unusual oleic, unsystematic Cis-9-octadecenoic as sources: the most common fatty acid in natural fats, particularly abundant in olive oil.

These phenomena also occur in nature, where the leptospira, which is chemoautotrophic, takes advantage of its carbon source and electron donor thanks to photosynthesis, which recycles nutrients by converting light energy into chemical energy, enabling it to survive in different environments [5].

The presence of certain chemicals in the environment can be harmful to bacterial DNA, which can undergo spontaneous degradation, modification of certain genes coding for enzymes crucial to bacteria, or biochemical modification through the addition of groups in the aquatic environment [25].

Deamination processes are frequently observed, leading to the transformation of cytosine into uracil, or methyl-cytosine into thymine. Similarly, alkyl groups such as methyl or ethyl can be added in an alkylation process. Finally, free oxygen radicals resulting from oxidative metabolism can cause hydroxylation, in particular generating hydroxy-guanine, which can pair erroneously with adenine, inducing G-C to A-T transversions.

Base analogs, such as the thymine analog 5-bromo-uracil, incorporate into DNA during replication and induce transitions during a subsequent replication cycle.

Intercalating agents, analogs of base pairs, such as ethidium bromide or acridine orange, insert themselves between bases, causing a shift in the reading frame.

In this way, nitrous acid induces deamination and nitrosoguanine causes alkylation.

All the above processes are responsible for gain of function, the appearance of new strains of leptospira bacteria that escape sanitary control, or loss of function [4, 5, 7, 18–22, 25].

Leptospira survive for several weeks, even months (up to 10 months) in warm, humid conditions (28–30°C), with neutral pH [15].

*The effect of humidity and oxygen tension* are linked. Very wet soils are hostile to the proliferation of many species of bacteria, including *Leptospira,* due to the reduced aeration of water-filled pores [22]. Soil flooding leads to the elimination of aerobic species. But, at the other extreme, soil drying significantly reduces metabolic activity and bacterial cell numbers.

Ultraviolet radiation (wavelengths 200–310 nm) is microbicidal because these wavelengths are absorbed by nucleic acids, damaging their structure and function. On the other hand, UV radiation is mutagenic because it generates pyrimidine dimers, which may be responsible for the virulence and reduced virulence of certain strains of leptospira observed in certain animals [25].

Leptospirosis is a worldwide bacterial disease. This disease is a zoonosis, meaning that it involves wild or domestic animals but can be transmitted to humans. Rats often like to live in damp areas close to houses, where household waste is disposed. These areas are a source of food for rats but also a favorable breeding ground for *Leptospir*a because rats and mice urinate and dispose of their feces where they take their meals [10–12, 17].

## 5. Conclusion

At the end of this chapter, leptospirosis remains a lesser-known disease in domestic, livestock and other environments. It is most often confused with typhoid fever, which occurs in the environment in the Democratic Republic of Congo and is little known anywhere else in the world.

However, our aim was to draw the attention of the general public, as well as the most at-risk individuals, such as livestock farmers and veterinarians, to the seriousness of leptospirosis.

This is because leptospirosis in the wild can interact with a wide range of products and waste from many different areas, giving rise to several different strains. However, the prevention of leptospirosis must begin in the environment.

To prevent leptospirosis, you need to:

- clean and disinfect livestock premises soiled by rodent urine and clean feed troughs;
- use protective equipment when handling livestock or their carcasses;
- Avoid swimming in freshwater rivers and do not walk barefoot;
- decontaminate urban areas and control the rodent population in rural areas.

## Acknowledgements

We would like to thank all those who kindly gave their insightful comments on our chapter, including the following teachers:

Matala érudit (colleague)

Joëlle kibulu (colleague)

We would like to thank Professor Tony Kayembe for his contribution to the improvement of the English language.

## Author details

Patrick Banza[1,2]

1 Higher Institute of Medical Techniques of Lubumbashi, Democratic Republic of Congo

2 Application Laboratory of the Higher Institute of Medical Techniques of Lubumbashi, Democratic Republic of Congo

*Address all correspondence to: patrickbanzakatolo@gmail.com

IntechOpen

© 2025 The Author(s). Licensee IntechOpen. This chapter is distributed under the terms of the Creative Commons Attribution License (http://creativecommons.org/licenses/by/3.0), which permits unrestricted use, distribution, and reproduction in any medium, provided the original work is properly cited.

## References

[1] Prescott ML, Harley PJ, Klein AD, Sherwood ML, Woolverton WS. Microbiology. 3 éme éd ed. Bruxelles: de Boeck university; 2010

[2] Sherwood LM, Willey JM, Woolverton CJ, Prescott LM. Microbiology of Prescott, 6 th éd, de Boeck Superior, Bruxelles; 2023

[3] Palka L. Microbiodiversity, Editions Materiology; 2018

[4] Tortora GJ, Funke BJ, Case CL. Introduction to Microbiology. 2eme éd ed. quebec: pedagogical renewal; 2012. pp. 15-16

[5] Madigan MT, Martinko JM. Biology of Microorganisms. 1eme éd ed. USA: Carbondale University of Southern Illinois; 2007

[6] World Health Organization. Geneva Genève Annual subscription/ Abonnement annuel Sw. fr./Fr. s. 346-02.2011 ISSN 0049-8114 Printed in Switzerlan

[7] Morand S and Colleagues. ANR project Futur. Health SEA; 2018-2021

[8] Ben A. Leptospira and Leptospirosis Current Topics in Microbiology and Immunology Book 387. springer; 2014

[9] Ahmed N, Devi SM, de los Valverde AM, Vijayachari P, Machang'u R, Ellis W, et al. Multilocus sequence typing method for identification and genotypic classification of pathogenic Leptospira species. Annals of Clinical Microbiology and Antimicrobials. 2006;**5**:28

[10] Blasdell KR et al. Association of rodent-borne. Leptospira spp. with urban environements in malaysian bornea. Plos Neglected Tropical Deseases. 2019;**13**(2):e 000714

[11] Brenner DJ, Kaufmann AF, Sulzer KR, Steigerwalt AG, Rogers FC, Weyant SR. Further determination of DNA relatedness between serogroups and serovars in the family Leptospiraceae with a proposal for Leptospira alexanderi sp. nov. and four new Leptospira genomospecies. International Journal of Systematic and Evolutionary Microbiology. 1999;**49**:839-858

[12] Bordes F et al. Habitat fragmentation alters the proprieties of a host-parasite network: Rodents and their helminths in south-east asia. Janimal Ecology. 2015;**84**:1253-1263

[13] Feresu SB, Steigerwalt AG, Brenner DJ. DNA relatedness of Leptospira strains isolated from beef cattle in Zimbabwe. International Journal of Systematic Bacteriology. 1999;**49**:1111-1117

[14] Holt J, Davis S, Leirs H. A model of leptospirosis infection in an African rodent to determine risk to humans: Seasonal fluctuations and the impact of rodent control. Acta Tropica. 2006;**99**:218-225

[15] Gorgé O, Lopez S, Hilaire V, Lisanti O, Ramisse V, Vergnaud G. Selection and validation of a multilocus variable-number tandem-repeat analysis panel for typing Shigella spp. Journal of Clinical Microbiology. 2008;**46**:1026-1036

[16] Ko AI, Goarant C, Picardeau M. Leptospira: The dawn of the molecular genetics era for an emerging zoonotic pathogen. Nature Reviews Microbiology. 2009;**7**:736-747

[17] Larry M, Bushet et Maria T, Pertejo V. Symptomatology of Leptospirosis. USA and Affiliates: MSD Manuel for Heatthcare Professionals; 2022

[18] Morand S and Colleagues. ANR Biodiversity project (ecology of rodent commuities and their pathogens in southeast Asia). health sea, Eco heath, projet CERO Path; 2008-2012

[19] Perez J, Brescia F, Becam J, Goarant C. Deciphering leptospirosis eco-epidemiology in New Caledonia. BMC Proceedings. 2011;**5**:P110

[20] Slack AT, Khairani-Bejo S, Symonds ML, Dohnt MF, Galloway RL, Steigerwalt AG, et al. Leptospira kmetyi sp. nov. isolé à partir d'une source environnementale en Malaisie. International Journal of Systematic and Evolutionary Microbiology. 2009;**59**:705-708

[21] Sonrier C, Branger C, Michel V, Ruvoën-Clouet N, Ganiere JP, Andrefontaine G. Evidence of cross-protection within Leptospira interrogans in an experimental model. Vaccine. 2000;**19**:86-94

[22] Thaipadungpanit J, Wuthiekanun V, Chierakul W, Smythe LD, Petkanchanapong W, Limpaiboon R, et al. Un clone dominant de Leptospira interrogans associé à une épidémie de leptospirose humaine en Thaïlande. PLoS Neglected Tropical Diseases. 2007;**1**:e56

[23] Wayne LG, Brenner DJ, Colwell RR, Grimont PAD, Kandler O, Krichevsky MI, et al. Report of the ad hoc committee on reconciliation of approaches to bacterial systematics. International Journal of Systematic Bacteriology. 1987;**37**:463-464

[24] DOJOZ. Summary of Ecology 8th Ed. Paris: Dunod; 2006

[25] Klug WS, Cummings MR, Charlotte. A spencer genetics. In: New Horizons, Pearson éducation. 8th ed. paris; 2006

## Chapter 7

# Leptospirosis: New Insights and Researches

*Farhang Babamahmoodi and Abdolreza Babamahmoodi*

### Abstract

Leptospirosis is a tropical re-emerging zoonotic infectious disease caused by spirochetes of the genus Leptospira. Leptospirosis is now seen as a major global public health issue, affecting both developing and industrialized nations. The economic impact and true incidence of the disease are not fully understood. It is estimated that tens of millions of individuals are affected by leptospirosis annually, with approximately 1,000,000 cases reported worldwide. The disease has a case-fatality rate ranging from 5 to 25%, resulting in 59,000 deaths and an estimated 2.9 million disability-adjusted life years lost each year, highlighting its substantial socioeconomic impact. Leptospirosis is increasingly affecting urban areas due to urbanization and proximity to disease-carrying animals. Challenges include a lack of awareness among healthcare professionals and the public, as well as climate change contributing to the spread of the disease. Addressing these challenges will require improved surveillance, public education, sanitation improvements, and research into new diagnostic tools and vaccines.

**Keywords:** leptospirosis, Weil's syndrome, re-emerging, diagnosis, neglected disease

### 1. Introduction

Leptospirosis is a zoonotic disease caused by *Leptospira* bacteria, which are gram-negative, aerobic, motile, helicoidal, and thin spirochetes of the genus *Leptospira.* The disease was first clinically reported in 1803 by William Wittmen, and later, in 1886, Weil described the jaundice component of the disease, leading to the disease being named after him [1].

Since 2018, the phylogeny of *Leptospira* has been extensively revised through large-scale whole-genome sequencing. Among the 64 recognized species, 60 have been formally documented following the regulations of the International Code of Nomenclature of Bacteria. These species are divided into two clades and four subclades. Subclade P1, which consists of 17 species, includes both newly identified species and those traditionally known to be pathogenic, such as *Leptospira interrogans, L. kirschneri, L. noguchii, Leptospira alexanderi, L. weilii, L. alstonii, L. borgpetersenii, L. santarosai, L. kmetyi*, and *Leptospira mayottensis.* Subclade P2 comprises 20 species, incorporating those previously considered to have intermediate or unclear pathogenicity. Subclades S1 and S2 encompass 22 and 5 species, respectively, including those previously classified as 'saprophytes' [2].



IntechOpen

Leptospirosis is a neglected tropical infectious disease caused by spirochetes of the genus *Leptospira*, and is considered a re-emerging zoonosis [3]. In recent years, there has been increasing recognition of leptospirosis as a significant global public health concern, impacting both developing and developed nations. The economic impact of this disease remains uncertain, but it is widely acknowledged that the true prevalence of leptospirosis is underestimated, particularly in endemic areas where the disease often goes undiagnosed. Leptospirosis is believed to affect millions of individuals annually, with an estimated 1,000,000 cases reported worldwide each year. The fatality rates for these cases range from 5 to 25%, resulting in approximately 59,000 deaths annually and causing an estimated loss of 2.9 million disability-adjusted life years per year, highlighting its substantial socioeconomic consequences [4]. Diagnosing leptospirosis clinically is difficult due to the presence of non-specific symptoms, often leading to misidentification with other illnesses like dengue fever, influenza, and hepatitis, which contributes to its underreporting. Additionally, around 90% of cases either show no symptoms or exhibit mild ones. The symptomatic cases can progress from a mild, self-limiting febrile illness to severe forms, characterized by multiple system complications that can result in a rapidly progressing, life-threatening condition [5]. Risk factors for leptospirosis encompass both behavioral and environmental elements [4]. Research in this field aims to evaluate the risk factors contributing to leptospirosis, describe the changing patterns of the disease, and identify alternative reservoirs of Leptospira that infect humans. Studies have also focused on identifying biological markers with predictive value and addressing major issues in leptospirosis research to move forward with adequate support [5]. This disease imposes an unknown economic burden and is remarkably underestimated and underdiagnosed in endemic regions [6]. The global research agenda for leptospirosis emphasizes the need for international cooperative efforts and the development of a multicenter prospective cohort study to understand the clinical manifestations and outcomes of leptospirosis. Data from prospective surveillance studies suggest that most human leptospiral infections in endemic areas are mild or asymptomatic, highlighting the need for continued research and surveillance to address this threat [5].

## 2. Epidemiology and global burden

*Geographical distribution:* Leptospirosis is a significant global issue. According to the World Health Organization (WHO), there is an endemic human leptospirosis rate of 5 cases per 100,000 individuals annually and an epidemic human leptospirosis rate of 14 cases per 100,000 people annually. However, recent assessments suggest that leptospirosis is currently recognized as the primary cause of morbidity and mortality in endemic reigon [7]. The prevalence and incidence of leptospirosis are affected by geographical and climatic conditions, including temperate and warm climates, the rate of rainfall and humidity, especially after heavy rainfall or flooding due to hurricanes, the presence of surface and stable water, and an increase in rodent populations [8]. Leptospirosis is an endemic zoonosis with a wide distribution, demonstrating various epidemiological patterns, including transmission in urban, rural, and sylvatic areas [8]. This disease is frequently observed in Asia, Oceania, and South America. However, there is an increasing number of cases being reported in Europe and North America [9]. Leptospirosis has the highest morbidity and mortality rates in resource-poor regions with tropical climates, such as parts of Sub-Saharan Africa, Latin America, the Caribbean, South and Southeast Asia, and Oceania. Vulnerable populations, including rural subsistence farmers and urban slum dwellers, are especially affected in these areas [4].

Leptospirosis is re-emerging as a zoonotic disease in developed countries, and its actual incidence is probably higher than reported. In the United States, the reported annual cases are relatively low, around 100–150, but it is believed to be underdiagnosed and underreported. A study from 1998 to 2009 found 0.6 hospitalizations per 1,000,000 people [10].

In 2021, the European Union and European Economic Area (EU/EEA) had a notification rate of 0.20 confirmed leptospirosis cases per 100,000 population. The rates of endemic human leptospirosis varied significantly, from 0.5 cases per 100,000 population in Europe to 95 cases per 100,000 population in Africa. France reported the highest number of confirmed cases (308) at a rate of 0.46 cases per 100,000 population. The highest rates of confirmed cases were found in 45–64-year-old males and 15–24-year-old females, with rates of 0.44 and 0.14 cases per 100,000 population, respectively. In 2020, the notification rate was 0.14 confirmed cases per 100,000 population in the EU/EEA. Globally, the annual incidence of leptospirosis is estimated to be between 14.8 and 350 cases per 100,000 population, with approximately 1 million cases and 60,000 deaths annually [11].

In China, the incidence of leptospirosis is concentrated in specific geographic areas, with the highest rates observed in the southern region of Yunnan Province. However, the level of spatial clustering has decreased over time [8]. In Seychelles, the occurrence of leptospirosis is influenced by various environmental factors, including low elevations, fragmented landscapes, proximity to urbanized areas, and increased precipitation, leading to spatial variations in its incidence [12]. Conversely, in Saint Lucia, there was no clear relationship between location and leptospirosis incidence, with only a weak correlation found between rainfall and cases [12]. In Denmark, leptospirosis is rare and non-endemic, with the highest annual incidence observed in 2017 [8, 13]. The prevalence of leptospirosis varies among Middle Eastern countries, with Iran, Turkey, and Egypt having more published literature on the disease compared to Lebanon, Kuwait, and Saudi Arabia. While the prevalence is generally similar across most Middle Eastern countries, there are variations among specific nations [7]. This suggests that further research or specific epidemiological studies for each country are necessary to determine the countries with the highest prevalence of neglected diseases in the world. These differences in leptospirosis rates across regions highlight the need for tailored prevention and control strategies to address the varying burden of leptospirosis in different areas.

*Sources of Infection:* Pathogenic *Leptospira* are frequently present in the environment and reside in the kidneys of different wild and domestic reservoir hosts. The life cycle of leptospires includes being excreted in urine, surviving in the environment, infecting a new host, and spreading to the kidneys via the glomerulus or peritubular capillaries. Once in the renal tubular lumen, they colonize the brush border of the proximal renal tubular epithelium, allowing for long-term urinary shedding without significant adverse effects on the host reservoir. This suggests that the relationship between leptospiral infection and the reservoir host can be considered commensal [14]. Small mammals are the main reservoirs of pathogenic *Leptospira*, and large herbivores are significant sources of infection. Leptospiral species have been found in numerous mammalian species, including bats and pinnipeds. Leptospires have also been discovered in poikilothermic animals such as frogs and toads, and although they may not be significant reservoirs of human infection, they could play a role in the circulation of leptospirosis in the environment. While only a few studies have reported the isolation of leptospires from amphibians, further research is needed to understand their role in maintaining leptospires in nature. Although leptospirosis is primarily a zoonosis, humans can serve as accidental hosts. Human-to-human transmission, while rare, can occur through sexual intercourse and lactation. Transplacental transmission may also occur during pregnancy, leading to abortion or stillbirth [15].

*Transmission:* Leptospirosis can enter the body through cuts, scrapes, or mucous membranes, such as the eyes, mouth, or genitals. The transmission can occur through direct contact with an infected animal or indirect exposure to soil or water contaminated with the urine of an infected animal. Individuals at risk of direct contact with potentially infected animals include hunters, abattoir workers, veterinarians, farm workers (especially those involved in dairy milking), trappers, animal shelter workers, scientists, and technologists who handle animals in laboratories or during fieldwork. The likelihood of infection depends on the prevalence of leptospirosis in the area and the frequency and extent of exposure. Wearing appropriate protective equipment like rubber boots, gloves, and eyewear can help prevent many of these infections. Given that workplace health and safety regulations often cover these infections, conducting local risk assessments and providing training is crucial [14, 16].

Leptospirosis is often linked to work, leisure, and personal activities, with indirect contact with water or soil contaminated with leptospires being the most common way it spreads. Jobs like working in sewers, military exercises, and farming in rainy tropical areas pose significant risks, especially for agricultural workers like rice field workers, taro farmers, banana farmers, sugarcane, and pineapple field harvesters. Recreational activities involving freshwater sports and adventure sports have also contributed to the rising cases of leptospirosis. Additionally, personal exposure, affecting millions of people in tropical regions, is worsened by poor sanitation and housing conditions. The combination of urbanization and climate change has been identified as a major risk factor for both the increased cases and frequency of leptospirosis outbreaks. Understanding the impact of human-induced changes on the spread of leptospirosis is crucial, emphasizing the need for interdisciplinary research to address this public health issue [7, 16].

*Global Burden of Disease:* Estimating the global impact of leptospirosis is difficult due to factors such as inadequate surveillance, insufficient diagnostic tests, and under-recognition. The WHO has been taking the lead in improving global estimates of illness and mortality by addressing these issues. The seriousness of the disease's prevalence and impact is particularly pronounced in economically disadvantaged tropical countries, where the burden of leptospirosis has been undervalued. Moreover, a more thorough understanding of the environmental phases of the disease cycle and its transmission routes is necessary to improve disease surveillance, diagnostics, and interventions [7, 17].

The Leptospirosis Epidemiology Reference Group (LERG) of the WHO was established to assess the worldwide impact of leptospirosis on illness, death, and Disability Adjusted Life Years (DALYs). Due to limited available data, the LERG conducted a thorough review of existing literature and developed a modeling approach. The estimated global burden of leptospirosis offers an opportunity to assess a health strategy for prevention and control. The global burden of leptospirosis was not calculated until 2015, and it is estimated that there are approximately 1.03 million cases annually, resulting in a total of approximately 2.90 million DALYs. This represents an incidence of 41.8 DALYs per 100,000 population per year [17]. The illness is found all over the world and can lead to life-threatening symptoms such as pulmonary hemorrhage syndrome. Most cases and fatalities from leptospirosis happen in tropical areas, with 73% of cases and deaths occurring between the Tropics of Cancer and Capricorn. A notable portion of the worldwide burden of cases and deaths related to leptospirosis affects males aged 20–49 years, with an estimated 48% of cases and 42% of deaths occurring in this age group [4].

The previous lack of recognition of leptospirosis as a significant zoonotic disease had hindered the consideration of a unified health approach to prevention and control. However, with the estimation of the global burden of leptospirosis, there is now an opportunity to assess a One Health strategy for prevention and control. Leptospirosis is a significant zoonotic disease that leads to illness and fatalities in humans, primarily affecting regions with substantial subsistence farming and pastoral communities. Further work is required to measure the economic impact of leptospirosis, including an evaluation of its effect on animal health. As climate change increases the frequency and severity of extreme weather events, efforts to address the disproportionate impact of leptospirosis have become increasingly important. Leptospirosis outbreaks often occur after floods and storms, highlighting the need to gain a better understanding of the environmental phase of the disease cycle for improved disease surveillance and interventions [3, 7].

Farmers, veterinarians, sewage workers, and outdoor recreational workers are more likely to get infected due to regular exposure to environments that may be contaminated. Among these workers, most infections result from direct contact with infected animals, with the highest risk associated with swine farming, slaughterhouses, and the meat industry. Risk factors for leptospiral infection include working overtime, coming into contact with animals during work, encountering rats at the workplace, and living less than 200 meters from a river. Wearing boots while working and washing hands with soap after work are recognized as protective factors [12, 14, 18].

Leptospirosis presents a socio-ecological challenge that frequently arises within the framework of social inequality. Consequently, it is crucial to assess and tackle the investment rationale for interventions that focus on the fundamental environmental conditions and infrastructure shortcomings, such as open sewers in urban slum communities, to achieve lasting advancements against this overlooked disease.

## 3. Pathology and pathogenesis

Pathogenic leptospires are commonly present in nature, and humans, as accidental hosts, typically show short-term urinary shedding of leptospires. In reservoir hosts, leptospires colonize the brush border of the proximal tubules for extended periods, without causing apparent adverse effects in the host but aiding in the spread of the microorganisms. However, during infections in humans and animals, the proximal tubules are the main sites of significant lesions, resulting in morphological and immunoenzymatic changes in the brush border, among other alterations. It is expected that the presence of microorganisms in the reservoir host may lead to subtle pathophysiological manifestations, at least in certain cases. In regions with high disease transmission, humans may exhibit clinical and serological characteristics similar to asymptomatic urinary carriers, resembling what is observed in reservoir hosts [14, 19].

Leptospirosis has two stages of pathogenesis. The first stage involves the penetration of tissue barriers into the body, which can occur through the skin or mucous membranes of the conjunctiva or oral cavity. Studies have shown that swallowing contaminated water is a risk factor for infection, highlighting the importance of oral mucosa as a gateway. The second stage involves hematogenous dissemination, wherein pathogenic leptospires enter the bloodstream and remain in the leptospiremia stage of the disease. During the first 8 days of fever, leptospiremia can be detected by quantitative PCR before the formation of antibodies and clearing of organisms from the bloodstream [20]. Leptospiral LPS (lipopolysaccharide) has a unique methylated phosphate residue not found in any other form of lipid A, and human TLR4 is unable to recognize it. Murine TLR4, on the other hand, can recognize leptospiral LPS, indicating that the mouse's innate immune response is adaptive to leptospiral infection. The absence of leptospiral lipid recognition in human cells, as opposed to murine cells, may result in the inability to generate appropriate responses, ultimately leading to susceptibility to leptospirosis [21].

Leptospirosis infection at high levels can trigger a robust immune response leading to severe conditions like sepsis or organ failure. Patients with severe cases often show increased levels of cytokines such as IL-6 and TNF-α. Elevated IL-6 and IL-10 are linked to higher mortality rates, potentially inhibiting protective immune responses. Non-specific T-cell activation may also play a role in the disease. A study on athletes exposed to contaminated water revealed HLA-DQ6 as a significant risk factor for leptospirosis, suggesting a superantigenic mechanism for susceptibility, although the specific antigens involved are unknown [14, 19, 22].

Leptospirosis primarily targets the liver, with pathology reports from fatal cases indicating sinus congestion and space of Disse dilatation. Studies have shown leptospires between hepatocytes, and recent research highlighted leptospiral penetration into the disc space and preferential binding to the perihepatocyte junction. Hepatocyte apoptosis is also observed. Damage to hepatocytes and intercellular junction disruption can lead to bile leakage into sinusoidal blood vessels, resulting in high direct bilirubin levels in icteric leptospirosis. Hemolysis induced by leptospirosis may occasionally elevate indirect bilirubin levels [23].

Pathological changes in the lungs, such as alveolar hemorrhage and pulmonary congestion, are frequently observed in severe cases of leptospirosis. In a study of fatal cases, all patients had pulmonary petechiae at the pleural level and 60% had severe bleeding at the cut lung levels. Histologically, bleeding was observed in the alveolar septum and intra-alveolar space. Immunohistochemistry of the lung tissue revealed granular material representing leptospiral antigen in macrophages in the septa and alveoli. The guinea pig model of leptospirosis mimics the pulmonary hemorrhage seen in humans, with studies showing significant immunoglobulin and complement deposition along the alveolar basement membrane. Pulmonary hemorrhage, a leading cause of death in leptospirosis in certain regions, may be linked to coagulation abnormalities common in severe cases. This pulmonary complication is not solely a lung issue but also reflects severe systemic disease, as evidenced by risk factors for severe pulmonary hemorrhagic syndrome (SPHS) including respiratory rate, hypokalemia, elevated serum creatinine, shock, and Glasgow Coma Scale score [14, 19, 24].

Renal complications in leptospirosis range from mild renal dysfunction to complete renal failure seen in Weil's syndrome. Severe cases may exhibit polyuria due to reduced expression of sodium-hydrogen exchanger 3, impacting sodium and fluid reabsorption in the proximal tubule. Kidney histological changes often involve tubular alterations and interstitial nephritis, with thinning or necrosis of tubular epithelium, tubular lumen dilation with casts, and cellular debris. In the reservoir host, tubules are crucial for leptospiral colonization. In humans, inflammatory responses triggered by leptospiral lipoproteins like LipL32 binding to TLR2 on renal tubular cells induce nitric oxide synthase and monocyte chemoattractant protein-1. This leads to tubular inflammation and interstitial nephritis characterized by cell infiltration. Interstitial nephritis typically worsens in the initial disease phase. Some patients may experience persistent renal issues with tubular atrophy and interstitial fibrosis on biopsy, but most recover normal renal function [14, 19, 25].

## 4. Clinical manifestations

Leptospirosis can range from a mild flu-like illness to a severe life-threatening condition, affecting different organ systems and leading to a variety of symptoms often confused with other causes of acute febrile syndrome. The incubation period typically lasts from 7 to 12 days but can range from 3 days to a month [26].

*General manifestations:* Patients with leptospirosis often experience sudden fever, chills, and headaches, which are non-specific symptoms also found in other febrile illnesses. Additional symptoms include muscle pain, tenderness in the calves and lower back, and conjunctival suffusion. Leptospirosis is typically distinguished by ocular findings, such as subconjunctival hemorrhage and icterus, and rashes are uncommon. An erythematous rash on the pretibial areas of both legs may indicate leptospirosis [3, 11, 14, 19].

Patients with leptospirosis commonly experience non-productive cough, gastrointestinal symptoms, and abdominal pain. Gastrointestinal symptoms may lead to dehydration, and abdominal pain may be due to acalculous cholecystitis and/or pancreatitis. Abnormal serum amylase and/or lipase levels are relatively common in hospitalized leptospirosis patients. Impaired renal function can elevate pancreatic enzyme levels. While most cases of pancreatitis due to leptospirosis are self-limiting, some are severe and fatal [27].

Severe leptospirosis involves dysfunction of several organs, including the liver, kidneys, lungs, and brain. The condition known as Weil's disease, characterized by jaundice and renal failure, was initially documented in 1886 and is still one of the most identifiable forms of leptospirosis. Signs of organ dysfunction signify a more advanced infection stage, and these symptoms can emerge suddenly and affect a significant portion of patients upon presentation [14, 27].

*Hepatic manifestations:* Patients suffering from leptospirosis generally display mild to moderate increases in liver transaminases and direct bilirubin levels. The occurrence of jaundice varies significantly across different cases, and acute hemolytic anemia can be a contributing factor. It is common for patients to experience leukocytosis and thrombocytopenia, but typically not to the degree that would result in spontaneous hemorrhaging. The presence of leukopenia in conjunction with thrombocytopenia and anemia suggests potential bone marrow suppression. Katz et al. discovered a strong correlation between infection and the Icterohaemorrhagiae serogroup, jaundice, and heightened bilirubin levels [28]. Occasionally, elevated levels of indirect bilirubin may occur due to hemolysis caused by leptospirosis [14, 19, 27].

*Hematologic manifestations:* Leptospirosis frequently results in clinical indications of bleeding, such as petechiae, ecchymosis, and epistaxis. While thrombocytopenia typically does not lead to spontaneous bleeding, certain patients may encounter significant gastrointestinal or pulmonary hemorrhage. Coagulation irregularities, including extended prothrombin time, are prevalent in severe cases of leptospirosis and are linked to severe bleeding. The exact cause of bleeding in leptospirosis is not entirely understood, but research suggests that it may involve ongoing fibrinolysis, coagulation activation, impaired anticoagulation, and thrombocytopenia. Platelet dysfunction has also been implicated in the occurrence of bleeding complications in leptospirosis [29].

*Renal manifestations:* Leptospirosis can lead to kidney complications, manifesting as elevated blood urea nitrogen and creatinine levels, pyuria, hematuria, and increased urine protein levels. A unique feature is the development of non-oliguric potassium wasting nephropathy, involving compromised sodium reabsorption and potassium loss. In instances of decreased oral intake due to symptoms like nausea and high-output renal failure, patients may face dehydration, increasing the risk of oliguria and renal failure. In regions lacking access to peritoneal or hemodialysis, renal failure can be a common cause of death in severe cases of leptospirosis [11, 14, 15, 30].

Leptospirosis is a frequent cause of acute kidney injury (AKI), with renal complications ranging from mild urinary irregularities to severe AKI necessitating dialysis. Acute tubulointerstitial nephritis (ATIN) is the most prevalent renal pathology associated with leptospirosis, and hypokalemia and sodium wasting are commonly observed in laboratory tests. AKI induced by leptospirosis typically does not involve reduced urine output and hypokalemia accounts for a substantial number of cases. Tubular dysfunction and related electrolyte imbalances, such as bicarbonaturia, glycosuria, and decreased sodium reabsorption in the proximal tubule, are also frequently seen in leptospirosis. Hyponatremia is a common occurrence in leptospirosis and is attributed to various factors, including increased urinary sodium loss and defects in Na+/K + -ATPase. The downregulation of the sodium-potassium-2-chloride co-transporter in the medullary thick ascending limb of the loop of Henle may contribute to the urinary loss of sodium and potassium in leptospirosis. During the initial stages of the disease, polyuria or non-oliguric acute kidney injury (AKI) can occur due to decreased expression of aquaporin 2, leading to a urinary concentration defect. As a compensatory mechanism, the expression of AQP2 increases during the recovery phase of AKI [14, 31].

Hypomagnesemia is a frequent occurrence in individuals with leptospirosis and acute kidney injury (AKI) due to the loss of magnesium caused by decreased NKCC2 on the apical membrane of the medullary thick ascending limb (mTAL). Monitoring of hypomagnesemia and hypophosphatemia resulting from increased excretion of magnesium and phosphate is crucial during leptospirosis infection. It is essential to carefully correct hypomagnesemia and hypophosphatemia, as rapid correction may inadvertently lead to elevated circulating magnesium levels due to reduced magnesium clearance in AKI caused by leptospirosis. Hypophosphatemia from proximal tubulopathy may contribute to metabolic encephalopathy and myopathy, emphasizing the need for regular assessment of serum magnesium and phosphate levels, particularly in severe cases of leptospirosis [31].

*Respiratory manifestations:* Patients with leptospirosis can exhibit primary pulmonary symptoms, such as cough, chest pain, breathlessness, and mild to severe hemoptysis, progressing to acute respiratory distress syndrome (ARDS). These pulmonary symptoms typically emerge between the fourth and sixth days of illness, and the disease progression can be rapid, leading to death in less than 72 hours. Dry cough, especially during the leptospiremic phase, is the most common pulmonary symptom. Severe pulmonary forms of leptospirosis (SPFL) may present as pulmonary hemorrhage, often massive, resulting in respiratory insufficiency and potential death from asphyxiation. Pulmonary hemorrhage can precede other manifestations of leptospirosis, including jaundice and acute renal failure. Hemoptysis, a sign of pulmonary hemorrhage, has been reported in 17–50% of patients. SPFL has a rapid and severe course, with high mortality rates ranging from 30 to 60%. In rare cases, patients may develop ARDS, requiring mechanical ventilation and associated with a mortality rate of up to 51%. ARDS is often linked to pulmonary hemorrhage due to endothelial damage. Myocarditis, renal failure, and/or overhydration in oliguria can lead to pulmonary edema. Upon general physical examination, patients with severe pulmonary leptospirosis may display tachypnea and cyanosis. Although jaundice is part of the classical triad, it may be inconspicuously absent in 30% of patients, and the severity of respiratory involvement does not necessarily correlate with the presence of icterus. The respiratory system examination may be entirely normal or reveal the presence of rales at the base. The most common cause of death in these cases is respiratory failure due to diffuse pulmonary hemorrhage [14, 32, 33].

*Nervous system manifestation:* Leptospirosis commonly presents with headache and fever, and approximately 50% of patients with suspected leptospiral meningitis have cerebrospinal fluid (CSF) pleocytosis, with most having normal CSF glucose and protein levels. Analysis of cerebrospinal fluid (CSF) in cases of aseptic meningitis commonly reveals a predominance of lymphocytes, with total cell counts reaching up to 500/mm3, protein levels ranging from 50 to 100 mg/mL, and normal glucose levels [34]. Leptospirosis can be a predominant cause of aseptic meningitis in certain areas. Patients with aseptic meningitis due to leptospirosis may not present with jaundice, making diagnosis more challenging. In severe cases, altered mental status may indicate meningoencephalitis, and other neurological complications such as hemiplegia, transverse myelitis, Guillain–Barré syndrome, and encephalitis can also occur [14, 35, 36].

*Ocular manifestation:* The frequency of ocular manifestations in leptospirosis ranges widely from 2 to 90% during the acute phase, with these signs typically appearing in the immune phase. These manifestations encompass various conditions such as chemosis, subconjunctival hemorrhage, cataracts, glaucoma, icteric sclera, pericorneal congestion, interstitial keratitis, and uveitis, which can affect one or both eyes. Uveitis, occurring in 40% of patients with systemic disease, can manifest as anterior or panuveitis, with the latter being the most severe form, involving non-granulomatous reactions, vitritis, periphlebitis, and other ocular complications. Fortunately, the prognosis for ocular findings in leptospirosis, including uveitis, is generally favorable with prompt treatment and accurate diagnosis [1]. One study suggests that non-granulomatous panuveitis, hypopyon, and vitreous infiltration are predictors of leptospiral uveitis in the absence of retinochoroiditis [37].

*Pregnancy and leptospirosis:* Pregnant women infected with leptospirosis may exhibit symptoms like fever, jaundice, and headache. The estimated incidence of leptospirosis in pregnancy is around 1.3 per 10,000 in women with fever or jaundice, potentially higher in endemic regions. Adverse fetal outcomes are more prevalent when the infection occurs during the second trimester compared to the third trimester. The clinical presentation of leptospirosis in pregnancy often overlaps with other pregnancy-related conditions such as Pre-Eclampsia, Acute Fatty Liver of Pregnancy, and HELLP Syndrome, making diagnosis challenging. In a significant number of cases, patient history revealed potential exposure to leptospirosis, emphasizing the importance of considering this infection in the differential diagnosis [38].

## 5. Prognosis and risk factors for morbidity and mortality

Leptospirosis is a serious and potentially fatal disease, with a mortality rate ranging from 0.87 to 2.49%. The prognosis of the illness hinges on factors such as disease severity, timely diagnosis, and immediate intensive treatment. Without proper intervention, complications like kidney failure, meningitis, liver damage, and respiratory distress can arise, leading to a case-fatality rate exceeding 50% in severe cases. Various risk factors linked to severe leptospirosis and mortality include age, clinical manifestations, and laboratory results. It is crucial to recognize that the transmission and clinical manifestations of leptospirosis can differ based on environmental and socioeconomic factors. Therefore, predictors of severe outcomes need to be assessed within each specific epidemiological context to account for regional peculiarities. Altered mental status, starting with confusion and obtundation without focal neurological signs, stands out as a significant independent predictor of a fatal outcome in leptospirosis. Other risk factors for death encompass hypotension, oliguria, age over 36 years, and respiratory insufficiency. Advanced age and lung involvement, indicated by dyspnea or alveolar infiltrates on chest X-rays, are correlated with higher mortality rates.

Moreover, various negative prognostic indicators identified in studies include acute renal failure, hypotension, arrhythmias, shock, multiple organ dysfunction, leukocytosis, anemia (hematocrit ≤30%), and a platelet count below 100,000/μL [14, 39].

Patients with leptospirosis generally have a good prognosis with appropriate supportive care. Those with acute renal failure necessitating dialysis often recover most of their renal function, although some may exhibit mild persistent renal impairment. Despite recovery, a significant portion of patients may encounter chronic post-leptospirosis symptoms. Around 30% of individuals report ongoing complaints following acute leptospirosis, which can manifest as fatigue, myalgia, malaise, headache, and weakness. Additionally, ocular complications such as uveitis can arise during the recovery phase of leptospirosis. This highlights the importance of continued monitoring and follow-up care for patients even after the acute phase of the illness has passed [14].

## 6. Diagnosis

Leptospirosis diagnosis involves detecting the organism or its components in body fluids, isolating Leptospira in cultures, or identifying specific antibodies. The choice of tests and specimens collected hinges on when symptoms started and how long they have been present [18].

*Molecular Diagnosis:* Leptospiral DNA can be amplified from various bodily fluids and organs using PCR and other assays, with quantitative PCR assays targeting different genes for diagnostic use. PCR detection in blood samples is more sensitive than culture; however, serology ultimately detects more cases. PCR assays currently can identify the infecting serovar, which has epidemiological and public health value [40]. Whole-genome sequencing has been applied to the diagnosis of neurological leptospirosis, and it is likely that direct serovar identification will be possible in the near future [41].

*Laboratory Isolation and Identification*: Leptospires are cultured using specialized media, and the timing of culture depends on the accurate onset of symptoms. Blood cultures should be obtained promptly after the patient's presentation, with a small amount of blood inoculated into a suitable medium. While multiple cultures increase recovery rates, this is often challenging. Other samples like cerebrospinal fluid (CSF) and peritoneal dialysate can be cultured in the first week of illness. Urine cultures are recommended from the second week onward, as urinary excretion can persist for weeks. Cultures are typically maintained at 28–30°C and examined weekly under dark-field microscopy for up to 13 weeks [14, 31].

*DOI: http://dx.doi.org/10.5772/intechopen.1008892*

Leptospires are identified through serological or molecular techniques, with sequence-based identification of Leptospira spp. becoming more common. For epidemiological purposes, strain subtyping can be done using methods like restriction enzymes or PCR variations to create distinct banding patterns. Modern techniques like MLVA and MLST offer reproducible and detailed sub-serovar-level information [42, 43].

Serological diagnosis is crucial for identifying leptospirosis due to limited culture and PCR capabilities. IgM antibodies typically appear in blood 5–7 days post-symptom onset. The Microscopic Agglutination Test (MAT) is the gold standard for serological diagnosis, involving live antigen reactions with patient sera to detect agglutination. MAT is serogroup-specific but may not pinpoint the infecting serovar accurately due to cross-reactions. Confirming a diagnosis requires paired sera showing a fourfold titer increase. While MAT is complex and requires live cultures, it is valuable for epidemiological studies. Other serological tests like ELISA are used for rapid screening, but MAT remains the preferred method for leptospirosis diagnosis despite its limitations [14, 18].

Due to the complexity of MAT, rapid screening for leptospiral antibodies in severe disease has been developed. IgM antibodies can be detected in the first week of illness, allowing diagnosis to be confirmed and treatment initiated, although this may be the most effective method. It has been repeatedly proven that the IgM test is more sensitive than MAT when the first sample is taken early in the course of the disease. Detection of IgM using ELISA is widely used, mainly using antigens prepared from culture of Lactobacillus Bifidum, although bacterial strain is also used. There are several products available. Recombinant antigens have also been used but have not been evaluated extensively. Recently, IgM testing has been developed into a variety of rapid tests intended for use in the laboratory without extensive testing or potency in the field. These include two types of test strips: latex agglutination, lateral flow, and dual-pass platform. However, the use of serological tests for early diagnosis has important limitations and evaluation of the second sample should be considered mandatory. It is also recommended to confirm the diagnosis quickly with the test used [14, 18].

## 7. Patients management

*Treatment:* Leptospirosis is a mild condition that usually resolves on its own, but early antibiotic treatment can prevent the disease from progressing to serious levels in patients. Diagnosis depends on the patient's risk, infection history, and symptoms, and rapid diagnosis has been developed. However, negative results should not be relied on to exclude early infection, and empirical treatment should be started as soon as the diagnosis is made. Leptospira strains are naturally resistant to many antibiotics, but there is little evidence of resistance to antibiotics used to treat leptospirosis. In addition to rapid diagnosis and appropriate antimicrobial therapy, regular monitoring of antibiotic resistance is important [44]. The treatment for severe leptospirosis includes the use of intravenous penicillin, ampicillin, ceftriaxone, or cefotaxime. Ceftriaxone is comparable to penicillin, has once-daily dosing, and can be administered intramuscularly. For outpatients with early disease, doxycycline or azithromycin is recommended, while pregnant women and children can be given azithromycin or amoxicillin. These recommendations are based on various studies and clinical experience, including research demonstrating the benefits of doxycycline therapy in shortening the duration of illness and preventing the shedding of organisms in the urine (**Table 1**) [11, 14].

| | Drug | Dose | Duration | Points |
|---|---|---|---|---|
| Out Patients | Doxycycline | 100 mg po Bid | 5–7 Days | |
| | Amoxicillin or Ampicillin | 500 mg po Tid or Qid | 7 Days | In pregnancy and children <8 years: 25–50 mg/kg/day po in 3 divided doses for 7 days |
| | Azithromycin | 1 gm po one dose and then 500 mg po for 2 days | 3 Days | In pregnancy and children <8 years: 10 mg/kg (max 500 mg) on day one and then 5 mg/kg/day for 3 days |
| Hospitalized | Penicillin G | 1.5 million units IV /Q6h | 7 Days | Children: 250,000–400,000 units/kg/day in 4–6 divided doses/7 days |
| | Cefotaxime | 1 gr IV /q6h | 7 Days | |
| | Doxycycline | 100 mg IV/ Bid | 7 Days | |
| | Ceftriaxone | 2 gm IV once daily | 7 Days | |
| Chemoprophylaxis | Doxycycline | 200 mg po | Once a week | |
| | Azithromycin | 500 mg | once a week | |

**Table 1.**
*Leptospirosis treatment or chemoprophylaxis.*

The treatment for severe leptospirosis involves the use of antibiotics and supportive therapy. This includes intravenous hydration, potassium supplementation, and, in cases of oliguric renal failure, peritoneal dialysis or hemodialysis, which can be life-saving if initiated early. Prompt initiation of daily dialysis in critically ill patients has been shown to reduce mortality. Additionally, patients with respiratory failure may benefit from low tidal volume ventilation to improve survival rates [10, 14, 27, 33].

Leptospires are susceptible to various antibiotics including β-lactams, macrolides, tetracyclines, fluoroquinolones [45], and streptomycin. However, determining susceptibility is challenging because of long incubation times, serum-containing media, and difficulties in quantifying growth accurately. Recent microdilution methods may facilitate the study of large numbers of isolates against different antimicrobial agents, potentially identifying new treatments for leptospirosis [45, 46].

According to a meta-analysis, there is no evidence that antibiotics are more effective than a placebo in treating leptospirosis, and no significant differences have been found in preventing symptomatic infection. Clinical trials with other antibiotics are needed to improve understanding of the treatment for this neglected disease [47].

A study in India found that levofloxacin demonstrated efficacy against leptospiral serovars, suggesting its potential for treating leptospirosis. Similarly, enrofloxacin showed success in treating bovine leptospirosis in Mexico. However, further research is needed to assess the effectiveness of fluoroquinolones in human cases. Prompt diagnosis and appropriate antimicrobial therapy are crucial in preventing chronic illness [48].

*Complications of treatment:* The Jarisch-Herxheimer Reaction (JHR) is a transient immunological phenomenon that can occur during treatment for leptospirosis that

*DOI: http://dx.doi.org/10.5772/intechopen.1008892*

can lead to severe symptoms including shaking chills, fever, headache, myalgia, and skin rashes, and in some cases, may cause complications such as respiratory distress syndrome, myocardial injury, and seizures. The reaction is caused by non-endotoxin pyrogen and spirochetal lipoproteins, and its mediation in relapsing fever involves pro-inflammatory cytokines such as TNF, IL-6, and IL-8. JHR typically occurs within 24 hours after antibiotic treatment and resolves a few hours later, but can also occur in the context of other spirochetal infections. In severe cases, it can lead to complications and even death, particularly in neonates and pregnant women. Ocular complications, such as ocular syphilis and neuroretinitis, have been reported in association with JHR, leading to vision impairment and other ocular issues. Ophthalmologists should be aware of these complications, especially when initiating antibiotic treatment in patients with syphilitic disease. Early diagnosis and appropriate treatment can reduce the mortality rate of severe JHR cases [49].

*Prevention:* Preventing leptospirosis involves understanding its epidemiology, transmission, and implementing protective measures. This includes reducing exposure, using protective clothing, and immunization. The disease is closely associated with poverty and poor housing, leading to exposure to rodent reservoirs. While rodent abatement efforts have short-term benefits, they pose risks to children and wildlife. Therefore, constructing rodent-proof housing and implementing flood control projects are crucial but challenging prevention measures. Identifying high-risk activities and using personal protective equipment are also important. Additionally, immunizing animals and treating infected ones are vital for preventing the disease in humans [3, 7, 14].

The use of oral doxycycline has been studied as an antibiotic for leptospirosis, and some evidence shows protection. However, the evidence is insufficient, and more powerful studies are needed to confirm the main effects of chemoprevention. Limited research suggests that chemoprophylaxis with doxycycline (200 mg orally per week), starting 1 to 2 days before exposure and continuing, may be effective in preventing infection, especially in adults with high risk and short-term exposure. However, the best time for leptospirosis chemoprophylaxis has not yet been determined and further research is needed to determine the best time for chemoprophylaxis [14, 47].

A study conducted in Iran demonstrated that a weekly dose of 500 mg of azithromycin or 200 mg of doxycycline can effectively prevent leptospirosis without any significant differences [50].

*Vaccination:* Vaccination with the leptospirosis vaccine is limited to high-risk groups and response to floods and epidemics. Reports from Japan, China, and Cuba indicate that vaccination leads to a reduction in the incidence of leptospirosis but also leads to social disruption and disease. Due to concerns of excessive effect on the host protein, growing a vaccine from Leptospira has been proposed in medicine, but this vaccine is not yet widely used. The vaccines are not fully effective but are the only universal vaccine authorized to control leptospirosis in humans and animals. However, it also has disadvantages such as lack of immunity, reactogenicity problems, and protection against most serotypes contained in the vaccine. Research on recombinant subunit vaccines has shown promising results, but they provide only partial protection against Leptospira infections. More research is needed to develop an effective leptospirosis vaccine [4, 14, 51].

The development of a killed whole-cell vaccine for leptospirosis by the Pasteur Institute led to a significant reduction in leptospirosis cases among Parisian sewer workers. The recommended vaccination protocol involves initial immunization, followed by two booster doses and reimmunization every 2 years. Recent reports have confirmed the safety and efficacy of the vaccine marketed as Spirolept™. The active component of killed whole-cell vaccines for leptospirosis is leptospiral LPS, a serovar-specific antigen that provides protection against homologous or closely related serovars. A trivalent vaccine has been developed to protect against unrelated serovars. Research is focused on developing leptospirosis vaccines with a low side-effect profile and long-lasting, cross-protective immunity by improving understanding of the leptospiral outer membrane [51].

## 8. New challenges

One of the new challenges related to leptospirosis is the increasing incidence of this disease in urban areas. As urbanization continues to expand, more people are living in close proximity to rodents and other animals that carry the bacteria responsible for leptospirosis. This increases the risk of exposure to disease, particularly in areas with poor sanitation and drainage systems.

Another challenge is the lack of awareness and understanding of leptospirosis among healthcare professionals and the general public. This can lead to delays in diagnosis and treatment, as well as missed opportunities for disease prevention and control.

Climate change also contributes to the spread of leptospirosis, as warmer temperatures and more frequent extreme weather events create favorable conditions for the bacteria to survive and spread.

Addressing these challenges will require a multi-faceted approach, including improved surveillance and reporting of leptospirosis cases, public education and awareness campaigns, and efforts to improve the sanitation and control of rodent populations in urban areas. Additionally, research into new diagnostic tools and vaccines for leptospirosis is needed to prevent and manage the disease better.

The important research topics concerning human leptospirosis encompass several critical areas. One key focus is exploring the role of the host immune response in the pathogenesis of leptospirosis, which holds significant implications for developing novel therapeutic strategies. Additionally, genomic and transcriptomic analyses of Leptospira spp. are crucial for understanding strain diversity, identifying virulence factors, and detecting emerging strains that impact public health.

Developing a rapid and accurate diagnostic test using CRISPR-Cas technology, along with other innovative diagnostic methods, would enhance early detection of leptospirosis, benefiting both human and veterinary medicine. Research into the effects of climate change on the incidence and distribution of leptospirosis is essential, as it may enable the prediction of environmental conditions conducive to outbreaks. Identifying novel drug targets through high-throughput screening and structural biology approaches can improve therapeutic outcomes. Moreover, examining the long-term effects and complications of leptospirosis is vital for understanding its impact on the development and growth of affected countries.

This research should also extend to the mechanisms of transmission in both urban and rural settings, bolstered by the development of effective vaccines, thereby advancing prevention measures against the disease. Collectively, these research efforts offer a comprehensive understanding of leptospirosis and form the foundation for improved interventions and control strategies.

*DOI: http://dx.doi.org/10.5772/intechopen.1008892*

## 9. Conclusions

Leptospirosis is a re-emerging zoonotic infection that poses significant challenges to human health, particularly in tropical regions, but also in other regions as well, and after heavy rainfall and floods. It is also becoming a new threat in developed countries and urban areas and requires increased awareness among clinicians. Diagnosis is challenging due to non-specific symptoms and inadequate laboratory test results. Medical management of severe cases requires a multidisciplinary approach, and further studies of Leptospira-host interactions are needed. Overall, the challenges of leptospirosis include its varied clinical presentation, the need for accurate diagnosis, and the importance of preventive measures to reduce exposure risk.

## Author details

Farhang Babamahmoodi* and Abdolreza Babamahmoodi
Antimicrobial Resistance Research Center, Communicable Diseases Institute, Mazandaran University of Medical Sciences, Sari, Islamic Republic of Iran

*Address all correspondence to: rmmsabm@gmail.com

IntechOpen

© 2025 The Author(s). Licensee IntechOpen. This chapter is distributed under the terms of the Creative Commons Attribution License (http://creativecommons.org/licenses/by/3.0), which permits unrestricted use, distribution, and reproduction in any medium, provided the original work is properly cited.

## References

[1] Arrieta-Bechara CE, Carrascal-Maldonado AY. Ocular leptospirosis: A review of current state of art of a neglected disease. Romanian Journal of Ophthalmology. 2022;**66**(4):282-288

[2] Kahneman JAC, Leptospirosis: Epidemiology, Clinical Manifestations, and Diagnosis. 2024.

[3] Bharti AR, Nally JE, Ricaldi JN, Matthias MA, Diaz MM, Lovett MA, et al. Leptospirosis: A zoonotic disease of global importance. The Lancet Infectious Diseases. 2003;**3**(12):757-771

[4] Costa F, Hagan JE, Calcagno J, Kane M, Torgerson P, Martinez-Silveira MS, et al. Global morbidity and mortality of leptospirosis: A systematic review. PLoS Neglected Tropical Diseases. 2015;**9**(9):e0003898

[5] Rex DAB, Chanderasekaran J, Rai AB, Phukan H, Sarma A, Prasad TSK, et al. Leptospira and leptospirosis: New systems science insights on proteome, posttranslational modifications, and pathogen-host interaction. OMICS. 2022;**26**(5):280-289

[6] Sanhueza JM, Baker MG, Benschop J, Collins-Emerson JM, Wilson PR, Heuer C. Estimation of the burden of leptospirosis in New Zealand. Zoonoses and Public Health. 2020;**67**(2):167-176

[7] Bradley EA, Lockaby G. Leptospirosis and the environment: A review and future directions. Pathogens. 16 Sep 2023;**12**(9):1167. DOI: 10.3390/pathogens12091167

[8] Dhewantara PW, Mamun AA, Zhang WY, Yin WW, Ding F, Guo D, et al. Geographical and temporal distribution of the residual clusters of human leptospirosis in China, 2005-2016. Scientific Reports. 2018;**8**(1):16650

[9] Abela-Ridder B, Sikkema R, Hartskeerl RA. Estimating the burden of human leptospirosis. International Journal of Antimicrobial Agents. 2010;**36**(Suppl 1):S5-S7

[10] Traxler RM, Callinan LS, Holman RC, Steiner C, Guerra MA. Leptospirosis-associated hospitalizations, United States, 1998-2009. Emerging Infectious Diseases. 2014;**20**(8):1273-1279

[11] Gizamba JM, Mugisha L. Leptospirosis in humans and selected animals in sub-Saharan Africa, 2014-2022: A systematic review and meta-analysis. BMC Infectious Diseases. 2023;**23**(1):649

[12] Browne ES, Callefe JLR, Jesus ERS, Zeppelini CG, Cremonese C, Costa F. A systematic review of the geographic distribution of pathogenic leptospira serovars in the Americas, 1930-2017. Anais da Academia Brasileira de Ciências. 2022;**94**(3):e20201026

[13] Harran E, Hilan C, Djelouadji Z, Ayral F. Epidemiology of Leptospirosis: The first literature review of the neglected disease in the middle east. Tropical Medicine and Infectious Diseases. 24 Sep 2022;**7**(10):260. DOI: 10.3390/tropicalmed7100260

[14] Haake DA, Levett PN. Leptospirosis in humans. Current Topics in Microbiology and Immunology. 2015;**387**:65-97

[15] Dos Santos LF, Guimaraes MF, de Souza GO, da Silva IWG, Santos JR,

Azevedo SS, et al. Seroepidemiological survey on leptospira spp. infection in wild and domestic mammals in two distinct areas of the semi-arid region of northeastern Brazil. Tropical Animal Health and Production. 2017;**49**(8):1715-1722

[16] Steneroden KK, Hill AE, Salman MD. Zoonotic disease awareness in animal shelter workers and volunteers and the effect of training. Zoonoses and Public Health. 2011;**58**(7):449-453

[17] Torgerson PR, Hagan JE, Costa F, Calcagno J, Kane M, Martinez-Silveira MS, et al. Global burden of leptospirosis: Estimated in terms of disability adjusted life years. PLoS Neglected Tropical Diseases. 2015;**9**(10):e0004122

[18] Schreier S, Doungchawee G, Chadsuthi S, Triampo D, Triampo W. Leptospirosis: Current situation and trends of specific laboratory tests. Expert Review of Clinical Immunology. 2013;**9**(3):263-280

[19] De Brito T, Silva A, Abreu PAE. Pathology and pathogenesis of human leptospirosis: A commented review. Revista do Instituto de Medicina Tropical de São Paulo. 2018;**60**:e23

[20] Agampodi SB, Matthias MA, Moreno AC, Vinetz JM. Utility of quantitative polymerase chain reaction in leptospirosis diagnosis: Association of level of leptospiremia and clinical manifestations in Sri Lanka. Clinical Infectious Diseases. 2012;**54**(9):1249-1255

[21] Que-Gewirth NL, Ribeiro AA, Kalb SR, Cotter RJ, Bulach DM, Adler B, et al. A methylated phosphate group and four amide-linked acyl chains in leptospira interrogans lipid a. the membrane anchor of an unusual lipopolysaccharide that activates TLR2. The Journal of Biological Chemistry. 2004;**279**(24):25420-25429

[22] Reis EA, Hagan JE, Ribeiro GS, Teixeira-Carvalho A, Martins-Filho OA, Montgomery RR, et al. Cytokine response signatures in disease progression and development of severe clinical outcomes for leptospirosis. PLoS Neglected Tropical Diseases. 2013;**7**(9):e2457

[23] Miyahara S, Saito M, Kanemaru T, Villanueva SY, Gloriani NG, Yoshida S. Destruction of the hepatocyte junction by intercellular invasion of leptospira causes jaundice in a hamster model of Weil's disease. International Journal of Experimental Pathology. 2014;**95**(4):271-281

[24] Marotto PC, Ko AI, Murta-Nascimento C, Seguro AC, Prado RR, Barbosa MC, et al. Early identification of leptospirosis-associated pulmonary hemorrhage syndrome by use of a validated prediction model. The Journal of Infection. 2010;**60**(3):218-223

[25] Araujo ER, Seguro AC, Spichler A, Magaldi AJ, Volpini RA, De Brito T. Acute kidney injury in human leptospirosis: An immunohistochemical study with pathophysiological correlation. Virchows Archiv. 2010;**456**(4):367-375

[26] Morgan J, Bornstein SL, Karpati AM, Bruce M, Bolin CA, Austin CC, et al. Outbreak of leptospirosis among triathlon participants and community residents in Springfield, Illinois, 1998. Clinical Infectious Diseases. 2002;**34**(12):1593-1599

[27] Herath NJ, Kamburapola CJ, Agampodi SB. Severe leptospirosis and pancreatitis; a case series from a leptospirosis outbreak in Anuradhapura district, Sri Lanka. BMC Infectious Diseases. 2016;**16**(1):644

[28] Chang ML, Yang CW, Chen JC, Ho YP, Pan MJ, Lin CH, et al. Disproportional exaggerated aspartate transaminase is a useful prognostic parameter in late leptospirosis. World Journal of Gastroenterology. 2005;**11**(35):5553-5556

[29] Wagenaar JF, Goris MG, Partiningrum DL, Isbandrio B, Hartskeerl RA, Brandjes DP, et al. Coagulation disorders in patients with severe leptospirosis are associated with severe bleeding and mortality. Tropical Medicine and International Health. 2010;**15**(2):152-159

[30] Ghasemian RSM, Makhlough A, Suraki-Azad MA. The course and outcome of renal failure due to human leptospirosis referred to a hospital in north of Iran; a follow-up study. Caspian Journal of Internal Medicine. 2016;**7**(1):7-12

[31] Chancharoenthana W, Leelahavanichkul A, Schultz MJ, Dondorp AM. Going micro in leptospirosis kidney disease. Cells. 16 Feb 2022;**11**(4):698. DOI: 10.3390/cells11040698

[32] Gulati S, Gulati A. Pulmonary manifestations of leptospirosis. Lung India. 2012;**29**(4):347-353

[33] Vandroux D, Chanareille P, Delmas B, Gauzere BA, Allou N, Raffray L, et al. Acute respiratory distress syndrome in leptospirosis. Journal of Critical Care. 2019;**51**:165-169

[34] Wang N, Han YH, Sung JY, Lee WS, Ou TY. Atypical leptospirosis: An overlooked cause of aseptic meningitis. BMC Research Notes. 2016;**9**:154

[35] Puca E, Majko J, Puca E, Qyra E, Gega A, Pipero P. Acute encephalitis as initial presentation of leptospirosis. Journal of Infection in Developing Countries. 2017;**11**(4):361-363

[36] Babamahmoodi F, Babamhmoodi A. Recovery from intracranial hemorrhage due to leptospirosis. Case Reports in Medicine. 2011;**2011**:504308

[37] Sivakumar R, Balakrishnan V, Gowri P, Visalakshi J. Leptospiral uveitis: Usefulness of clinical signs as diagnostic predictors. Ocular Immunology and Inflammation. 2018;**26**(4):569-576

[38] Selvarajah S, Ran S, Roberts NW, Nair M. Leptospirosis in pregnancy: A systematic review. PLoS Neglected Tropical Diseases. 2021;**15**(9):e0009747

[39] Pongpan S, Thanatrakolsri P, Vittaporn S, Khamnuan P, Daraswang P. Prognostic factors for leptospirosis infection severity. Tropical Medicine and Infectious Disease. 11 Feb 2023;**8**(2):112. DOI: 10.3390/tropicalmed8020112

[40] Mazzotta E, Ceglie L, Giurisato I, Bellinati L, Lucchese L, Marchione S, et al. Persistence of leptospira borgpetersenii Serovar Hardjo in refrigerated raw milk: A transmission risk of leptospirosis to humans. Pathogens. 3 Mar 2021;**10**(3):291. DOI: 10.3390/pathogens10030291

[41] Lata KS, Vaghasia V, Bhairappanavar SB, Kumar S, Ayachit G, Patel S, et al. Whole genome sequencing and de novo assembly of three virulent Indian isolates of leptospira. Infection, Genetics and Evolution. 2020;**85**:104579

[42] Pavan ME, Cairo F, Brihuega B, Samartino L. Multiple-locus variable-number tandem repeat analysis (MLVA) of leptospira interrogans serovar Pomona from Argentina reveals four new genotypes. Comparative Immunology, Microbiology and Infectious Diseases. 2008;**31**(1):37-45

[43] Ahmed A, Thaipadungpanit J, Boonsilp S, Wuthiekanun V, Nalam K, Spratt BG, et al. Comparison of two multilocus sequence based genotyping schemes for leptospira species. PLoS Neglected Tropical Diseases. 2011;**5**(11):e1374

[44] Trott DJ, Abraham S, Adler B. Antimicrobial resistance in leptospira, brucella, and other rarely investigated veterinary and zoonotic pathogens. Microbiology Spectrum. Jul 2018;**6**(4). DOI: 10.1128/microbiolspec.arba-0029-2017

[45] Griffith ME, Moon JE, Johnson EN, Clark KP, Hawley JS, Hospenthal DR, et al. Efficacy of fluoroquinolones against leptospira interrogans in a hamster model. Antimicrobial Agents and Chemotherapy. 2007;**51**(7):2615-2617

[46] Ressner RA, Griffith ME, Beckius ML, Pimentel G, Miller RS, Mende K, et al. Antimicrobial susceptibilities of geographically diverse clinical human isolates of leptospira. Antimicrobial Agents and Chemotherapy. 2008;**52**(8):2750-2754

[47] Guzman, Perez M, Blanch Sancho JJ, Segura Luque JC, Mateos Rodriguez F, Martinez Alfaro E, et al. Current evidence on the antimicrobial treatment and chemoprophylaxis of human leptospirosis: A meta-analysis. Pathogens. 2 Sep 2021;**10**(9):1125. DOI: 10.3390/pathogens10091125

[48] Trifiletti RR, Lachman HM, Manusama O, Zheng D, Spalice A, Chiurazzi P, et al. Identification of ultra-rare genetic variants in pediatric acute onset neuropsychiatric syndrome (PANS) by exome and whole genome sequencing. Scientific Reports. 30 Jun 2022;**12**(1):11106. DOI: 10.1038/s41598-022-15279-3

[49] Butler T. The Jarisch-Herxheimer reaction after antibiotic treatment of spirochetal infections: A review of recent cases and our understanding of pathogenesis. The American Journal of Tropical Medicine and Hygiene. 2017;**96**(1):46-52

[50] Alikhani A, Salehifar E, Zameni F, Rafiei A, Yazdani-Charati J, Delavaryan L, et al. Comparison of azithromycin vs doxycycline prophylaxis in leptospirosis, a randomized double blind placebo-controlled trial. Journal of Infection in Developing Countries. 2018;**12**(11):991-995

[51] Wang Z, Jin L, Wegrzyn A. Leptospirosis vaccines. Microbial Cell Factories. 2007;**6**:39